For Michelle

DINOSAURS
of the west

RON STEWART

Lone Pine Publishing
Edmonton, Alberta

Mountain Press Publishing Co.
Missoula, Montana

QE
862
.D5
S715
1988

Copyright © 1988 by Lone Pine Publishing and Ron Stewart

First Printed 1988 5 4 3 2 1

All rights reserved

No part of this book may be reproduced in any form without permission in writing from the publishers, except by a reviewer who may quote brief passages in a magazine or newspaper.

The Publishers:
Lone Pine Publishing
414, 10357 - 109 Street
Edmonton, Alberta
T5J 1N3

Typesetting by Pièce de Résistance Ltée, Edmonton, Alberta
Printing by The Jasper Printing Group Ltd., Edmonton, Alberta

Canadian Cataloguing in Publication Data

Stewart, Ron, 1951-
    Dinosaurs of the West
    ISBN 0-919433-37-5

    1. Dinosaurs - Juvenile literature.   I. Title.
QE862.D5S84 1988   j567.9'1   C88-091110-7

Library of Congress Cataloguing-in-Publication Data

Stewart, Ron, 1951-
    Dinosaurs of the West

  Summary: Examines the history and characteristics of dinosaurs in the western areas of Canada and the United States as revealed by the discovery of bones and other petrified evidence. Also discusses how scientists gather and identify dinosaur remains and includes a glossary of terms.

    1. Dinosaurs - West (U.S.) - Juvenile literature.
    2. Dinosaurs - Canada, Western - Juvenile literature.
    [1. Dinosaurs - West (U.S.) 2. Dinosaurs - Canada, Western]   I. Title.
QE862.D5S715 1988   567.9'1'0978   88-9013
ISBN 0-87842-225-0 (pbk.)

Design: Yuet C. Chan
Colour Illustration: Ewa Pluciennik
Black & White Illustration: Tenny Whitfield
Photography: Ron Stewart, unless credited otherwise.

Publisher's Acknowledgement

The publisher gratefully acknowledges the assistance of the Canadian Department of Communication, Alberta Culture, the Canada Council, the Alberta Foundation for the Literary Arts and the Recreation, Parks and Wildlife Foundation of Alberta in the production of this book.

Content

Acknowledgement 6
Preface 7

Part I The Lost World 9

The Hidden Treasure *10*
What Were the Dinosaurs *12*
What Is A Fossil? *13*
Geological Time *21*

Part II The Bestiary 35

An Introduction to the Dinosaurs *36*
The Dinosaurs *39*
Archaeopteryx *89*
Pterosaurs *92*
Marine Reptiles *94*

Part III The Mystery Continues 97

Dinosaurs Eggs and Young *98*
Warm-Blooded or Cold-Blooded? *102*
Death of the Dinosaurs *105*
Plants of the Cretaceous *112*
How Old Is It? *118*
Dinosaur Provincial Park *120*
Pioneer Dinosaur Hunters in Canada *122*
Field Collecting *128*
Recent Dinosaur Find *133*
The Role of the Dedicated Amateur *134*

Pronunciation of Dinosaur Names 137
Glossary 138
Museums 142

Acknowledgements

Sincere thanks are due to the many people who gave generously of their time and abilities, helping with the multitude of details necessary to see this book to its completion.

I especially wish to thank my wife, Michelle, without whose constant support and encouragement this project would never have reached completion.

I would like to thank the Department of Geology, University of Alberta, for its support. Particular thanks are extended to: Dr. N. W. Rutter, Chairman; Mr. Garth Milvain; Dr. C. R. Stelck, for his comments and suggestions; Dr. R. C. Fox, for his suggestions and critical comments; Dr. B. D. E. Chatterton; Dr. J. F. Lerbekmo; Dr. J. Krupicka; Mr. A. Lindoe; Dr. S. Speyer; Dr. B. Blackwell; Mrs. Gaileen Lyons; Mr. P. Black; Judy A. Powell, John Allan Library; Mr. Scott Reed; Mr. F. Dimitrov; Mr. G. Braybrook, Department of Entomology; Mr. D. Hildebrandt, Department of Entomology; Christina Barker, Mining Engineering; Mr. Randy Mandryk, Department of Zoology; Dr. R. A. Stocky, Department of Botany; the staff of the University Archives.

Thanks are due to the staff of the Tyrrell Museum of Paleontology, Drumheller; notably to Dr. E. Koster, Director; Dr. Phillip Currie; Lynn Thornton, Publicity, and Collin Orthner. My gratitude to the staff of the Provincial Museum and Archives, Edmonton; Dr. J. Burns; Dr. R. Mussieux. Thanks to Dr. T. Jerzykiewicz of the I.S.P.G. in Calgary and to Dr.R.Harrington of the National Museum, Ottawa, Dr. J.W. Schopf, University of California, Dr. Elliott Burden, Memorial University, University of Toronto.

Special thanks to publisher Grant Kennedy, Mr. Raymond Gariepy and the staff of Lone Pine Publishing, who made this book possible.

I would also like to thank Miss Michelle Tatton; Master Gordon Hagan; Master Bobby Desnoyers; Miss Margaret Ann Hanrahan and her family; Kirk and Wendy Stewart of Stettler; Terry and Karl Berg; Mr. M Weiss; Mr. R. Yaciuk and Mr. Lester Foster of Hub Photo; Mr. Del Walters of Bedrock Lapidary. Special thanks to the long-distance operators of Alberta Government

Telephones, who were of great help in placing overseas calls. Finally, many thanks to my parents and family for their support, comments and encouragement, and to my grandmother, Mrs. Mary Tatton, who always believed in me.

To all those whom I have neglected to mention, my very special thanks. Without you, this book would not have been possible.

Preface

When scientists attempt to reconstruct the life of the dinosaurs, they may have only fossilized bone and teeth, perhaps impressions of the skin, or footprints preserved in mud. While this information allows paleontologists to reconstruct the skeletons of dinosaurs, it does not allow questions of dinosaur behaviour to be answered with any certainty. Paleontologists are always searching for better evidence to add missing pieces to the puzzle of the dinosaurs. Old theories are discarded as new ideas emerge. While this book will function as a guide in your understanding of the dinosaurs, like all books, it is static and remains fixed, while the scientific study of dinosaurs is always moving forward. The material presented in this book is drawn from a large number of sources, and while there may be some compelling arguments for some of these theories, it must be remembered that not all scientists would agree with the various interpretations given.

PART I

The Lost World

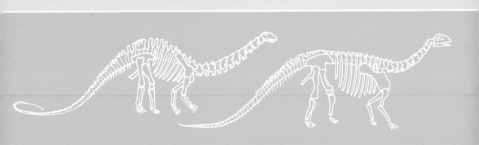

The Hidden Treasure

The ancient rocks of western North America hold the secrets to the Age of Dinosaurs. It is hard to imagine when one stands today on the empty prairie that this land was once a vast coastal plan, populated with strange tropical plants, and home to the dinosaurs.

Today, the prairies bear witness to the passage of the glaciers which moved across the continent. These great sheets of ice, in places several kilometres thick, changed the shape of the land itself, levelling hills and blanketing the countryside with glacial boulders and till (crushed rock and soil) which today hides the prairie's distant past. While the last glacier retreated from the plains about 10,000 years ago, this span of time is insignificant when compared with the 65 million years which have elapsed since dinosaurs roamed the earth.

To visit this ancient land of the dinosaurs, we do not need a time machine, for scattered amid the badlands and river valleys which scar the prairie we find the broken bones of these long-dead giants. The rivers have provided us with these windows into the past, by cutting down through the overlying glacial sediments, exposing the muds and sands which covered the land during the great Age of the Dinosaurs. When we leave the high level of the plains and descend into a river valley or a coulee, we are descending into time.

Let us imagine for a moment that we could take a journey back 75 million years and hang suspended in space, overlooking these plains. The first thing we would see is that the shape of the North American continent is far different than it is today. Much of it was underwater, covered by a shallow sea which extended from the Gulf of Mexico to the Arctic Ocean. Over time, the levels of the sea changed, alternately flooding and retreating from the continent. These periodic fluctuations of sea-level corresponded with the geological changes which were shaping the continent. It was as a direct result of these geological processes that the Rocky Mountains had begun to rise.

Rains falling on the eastern slopes of the Rockies over millions of years carried vast quantities of sediments out over a broad flood plain extending toward the east. These waters formed rivers which

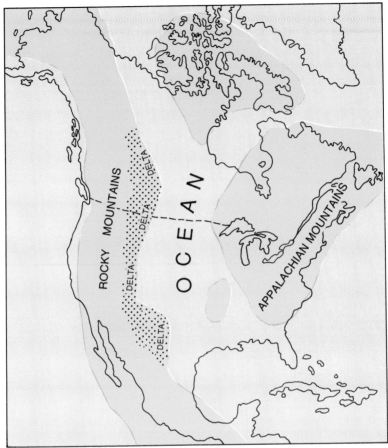

Approximate extent of ocean (75 million years ago).

carried the silts and sands, depositing them in great deltas which grew along the edge of the retreating sea, east of what is now the Alberta-Saskatchewan border. It was here, in the summer of 1874, that the first dinosaur remains were discovered in Canada in the valley of Rock Creek, N.W.T. (Morgan Creek, Saskatchewan).

The modern-day Dinosaur Provincial Park lies about 100 kilometres from that ancient sea. The landscape of the delta was similar to that of the modern coastal marshes of the southern United States. It was a land of cattail swamps and groves of trees growing on the sand and gravel levees created as the rivers meandered across the land. The wildlife inhabiting this country ranged from dinosaurs, turtles, crocodiles and other reptiles to primitive mammals which scurried through the undergrowth. Like modern swamps, this long-ago marsh was plagued with clouds of mosquitoes and blackflies which swarmed amid the dark conifer trees rimming the pools of water. Today, we find their fossils in the bits of amber (fossilized plant resin) which can be found in the coal seams which outcrop in Dinosaur Provincial Park.

Over millions of years, the diverse life of this ancient plain was trapped and preserved by the sediments which shaped the land. From the delicate insects in amber, the crumbling bones of the dinosaurs, to the seams of coal which first brought settlers into the badlands of Drumheller, each left some record of its passing, and each has a story to tell us.

Our search for the dinosaurs will take us from recent finds in Canada's Arctic to discoveries in British Columbia, Saskatchewan, Alberta and the western United States. The ancient sediments which today underlie western North America provide paleontologists (scientists who study fossils) with a glimpse into that distant time, when dinosaurs ruled the earth.

What Were The Dinosaurs?

The dinosaurs were reptiles which existed approximately 225 million years ago and ruled the earth unchallenged before becoming extinct about 65 million years ago.

When we think of dinosaurs, we think of giant creatures living in an ancient time when the world was very different from ours today. We think of great beasts like Apatosaurus (Brontosaurus), weighing more than twenty full-grown elephants. Other dinosaurs were tiny creatures like Ornitholestes, who scurried through the jungles. Still others were the stuff of our nightmares: great killers like the Tyrannosaurus rex, who stood over 6 meters in height and was 15 meters long from the tip of its tail to its teeth. These creatures weighed as much as 6.5 tonnes and stalked the jungles in search of food.

They were unique, these ancient dragons out of the past. They fill us with wonder when we touch their giant bones. The world will never see their like again.

Ornitholestes (bird robber). Small, primitive dinosaur from the Late Jurassic. It is thought that it may have captured small lizards and other small amimals. Length 2.1 m (6.5 ft.). (Cast). Tyrrell Museum.

What Is A Fossil?

Before we can understand dinosaurs, we must first explain what a fossil is. In simple terms, a fossil is the remains of a once-living plant or animal and includes casts, footprints and frozen remains. Although there are many different types of fossils, only a very few plants or animals fossilize. As a result, fossils are rare.

Fossil bone may look the same as original bone, but it may be very different. Most dinosaur bones are much harder than the original bone. How can this happen? What can change bone to hard rock?

Mineralization or Petrification

If an animal dies and is covered with sediment (mud or sand), there is a better chance it might become a fossil. If it is buried quickly, the body is protected by the soft mud or sand. The flesh of the animal rots (due to the action of bacteria in the sediments) leaving behind the bones or hard parts, which are covered and surrounded by sediment. This is just the beginning of the process of fossilization.

The process which changes original bone or wood to hard rock is called mineralization or petrification. Petrification is caused by groundwater carrying dissolved minerals through the soil. Two commonly found minerals are calcium carbonate or limestone — frequently found in deposits in water pipes or kettles; and, silica or quartz — found as grains of ordinary sand. Fossils are often

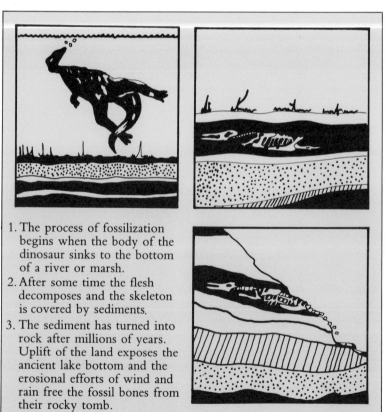

1. The process of fossilization begins when the body of the dinosaur sinks to the bottom of a river or marsh.
2. After some time the flesh decomposes and the skeleton is covered by sediments.
3. The sediment has turned into rock after millions of years. Uplift of the land exposes the ancient lake bottom and the erosional efforts of wind and rain free the fossil bones from their rocky tomb.

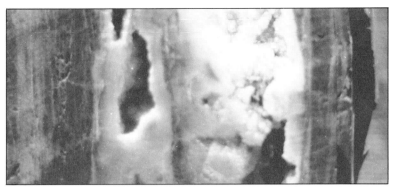

Dinosaur bone showing minerals (silica) filling an internal cavity. Tyrrell Museum.

composed of either calcium carbonate or silica.

If bone or wood is buried, the mineral bearing groundwater gradually replaces the original wood or bone. When we find petrified wood or bone, we are finding not animal or plant remains, but merely copies composed of minerals. Sometimes the copy is very good, allowing us to observe cell structure. Under different conditions, the copy only provides an approximate shape of the original material.

Casts and Molds

Sometimes when a plant or animal dies and is buried in sediment, the groundwater completely dissolves the remains. When this occurs, a hollow space is left in the sediments where the original was buried. If this hollow space is filled with minerals or mud, it will produce a fossil called a cast.

A cast is merely the external form of the fossil. While a cast may be a perfect copy of the external form of the fossil, it has no internal detail, unlike something which has been petrified. If we slice through a cast, all we find are solid minerals or mud. The fine details of the fossil have not been preserved.

On the other hand, if the sediments harden and become rock, and the hollow space is not filled with minerals, we have a mold of the buried plant or animal. This is a cavity in the rock which looks like the outside form of the original. There are no internal details in a mold, because all which remains of the original is just a hollow space.

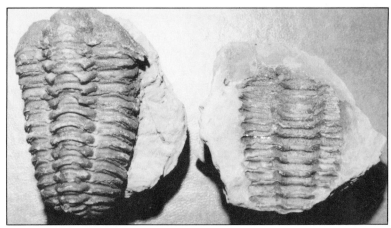

Trilobite cast and mold.

Plant impression.

Carbonization (Plant Impressions, Insect Impressions)

Carbonization is a process of fossilization which is quite common in producing impressions of fossil plants, and less common in fossil insects. After the remains are buried in fine sediments, anaerobic bacteria (bacteria requiring no air) break down the fossil material, releasing gases and leaving a fine carbon deposit on the rock.

In Toto (Total Preservation)

In toto or total preservation is very rare. This type of fossilization occurs in three ways: preservation in amber, preservation in ice and preservation in petroleum or bitumen.

Insect in amber.

Preservation in Amber

Amber is the fossilized resin or sap which once flowed out of ancient plants. Sometimes insects were attracted to the resin and became entrapped. If the plant sap fossilized as amber, the trapped insects would also become fossils. Unlike other forms of fossilization, where the original remains are replaced by minerals, the fossils in the amber are the actual insects. A fly in amber has not been replaced with minerals. The body and internal organs of the insects are still intact. If we slice into a piece of amber, we can examine the fossil insect under the microscope. The cell structure, and even the DNA of the insects, are preserved.

Preservation in Ice or Permafrost

The rarest fossils are those found frozen in ice or permafrost. In the Arctic, the ground remains frozen all year round, keeping any fossils buried in a permanent deepfreeze. The remains of baby mammoths have been found frozen in ice or mud, exactly as they were buried 30,000 years ago. While these fossils usually show evidence of some decay or deterioration, the cells and organs of these fossils are reasonably intact. It is thought by some scientists that one day it may be possible to clone a mammoth from its frozen cells.

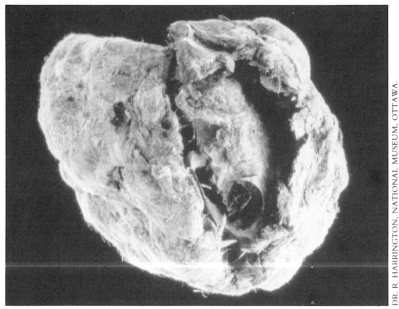

Ground squirrel preserved in permafrost. Approximate age 47,000 years.

Preservation in Oil or Asphalt

Another form of preservation is found in the Athabasca Tar Sands of northern Alberta, where fossil wood or plant materials are found preserved by petroleum or bitumen. These fossil remains have not been replaced by minerals but still contain original cellular material. Other examples of preservation by petroleum or bitumen are the tarpits at Rancho La Brea in Los Angeles, California. Here, fossil animals were trapped in sticky asphalt as natural oil evaporated. A large number of fossils found in this deposit consist of remains of mammoth, bison, horse, dire-wolves and sabre-toothed tigers. The fossils of vultures, eagles and other birds of prey are very common. It seems that the large number of animals trapped in the asphalt attracted scavengers, who also were frequently mired in the pits.

Trace Fossils

Trace fossils reveal the former presence or activities of organisms, but are not fossils of the organisms themselves. Trace fossils can

Hadrosaur trackways from the Peace River Canyon, British Columbia. Dinosaur remains in British Columbia are quite uncommon. This unique site contains more than 1,700 footprints. Extensive excavations were conducted before the area was flooded by a dam in the late 1970's.

Gastroliths or "stomach stones", probably from dinosaur. Collection U of A.

Trace fossil; track of trilobite (Middle Cambrian).

be such things as coprolites (fossilized animal dung or droppings), gastroliths (stomach stones), footprints or trackways and fossilized burrows in mud, produced by worms and other organisms. Trace fossils such as coprolites provide information about diet, while trackways and footprints can tell scientists how an animal moved. Information such as this can determine if the organism moved on two legs or four, how heavy the animal was and if it grouped together into herds. The study of trace fossils is a specialized sub-branch of paleontology and is called ichnology.

Geological Time

To help us understand the processes of the earth which gave rise to life, we must journey backward in time to the formation of our planet, about 4.6 billion years ago. Because little direct evidence exists for these beginnings, scientists have to formulate theories which explain the formation of the oceans, the atmosphere, and even life itself. It must be remembered that this chain of events is what some scientists think took place. Not all will agree with this interpretation.

Originally, the surface of the newborn planet was little more than barren rock. Over many millions of years, volcanoes poured out molten lava and hot gases. From these beginnings, our atmosphere, oceans and life itself were born.

For several hundred million years, these gases, which are thought to have contained nitrogen, carbon dioxide, sulphur compounds and water vapour, poured forth from the crust of the earth to form a primitive atmosphere. Rains fell during these millions of years, but each time the surface rocks were too hot for surface water to exist, and the rains boiled away into the sky as steam.

Basaltic lava, island of Hawaii, showing surface texture. The early surface of the Earth may have looked like this.

Over time, however, the rocks cooled and the waters filled the low-lying basins to become the first oceans. The rains dissolved minerals from the surface rocks to turn the oceans salty and to provide chemicals in solution for the greatest product of the new planet Earth: life.

Scientists are not sure how life was first formed. One school of thought places this chain of events in the oceans. In some tidal pool or on a beach, simple compounds of water, carbon dioxide and ammonia were united, perhaps by a stray bolt of lightning, to form amino acids, the first primitive proteins. Other scientists are of the opinion that life evolved on dry land, using clay minerals to form templates for the formation of simple organic proteins.

In 1953, Dr. S. L. Miller performed a now-classic experiment. He approximated the early atmosphere of the Earth in a flask and ran an electric spark into this mixture to duplicate lightning. When the contents of the flask were analyzed after several days, amino acids were found in the solutions. These same amino acids are found today in all living organisms. They contribute to the complex molecules of DNA which control all life on the planet.

Scientists need to be precise when speaking in terms of time. To help in this discussion, it is necessary that we begin to use some of these terms. The period of earth's history from the formation of the planet, 4.6 billion years ago, until approximately 600 million years ago is known as the Precambrian Era. This is an

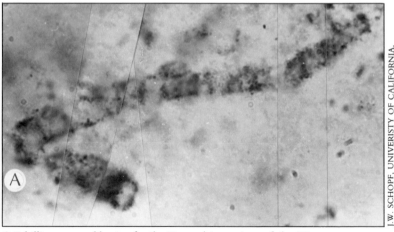

3.5 billion years old microfossils. From thin sections of the Warrawoona Group, North West Australia.

Stromatolite mounds from the north shore of Great Slave Lake (about 2.3 billion years old).

immense span of time which represents about 85 percent of Earth's history.

For most of the Precambrian period, fossils are rare and microscopic. The only large structures found containing remains of fossil life are stromatolite mounds. These organisms make limestone reefs, somewhat like modern coral reefs, which were produced as tiny blue-green algae converted carbon dioxide to simple sugars using light. This latter process is called photosynthesis, a very important development of life.

The waste product of photosynthesis is oxygen. It is this waste product of plants, nearly all of which use photosynthesis, which supports all higher animal life forms. The oxygen you are breathing as you read this book was produced by a plant. The molecules of oxygen might have come from the Precambrian algae which built the stromatolite mounds or from a great fern which grew in a coal swamp before the time of the dinosaurs.

But to return to our blue-green algae, it was as a result of these early plants that levels of oxygen in the ocean began to rise. One billion years after the formation of the Earth, life had become so abundant that its waste products had begun to alter the chemistry of the planet. This increase in oxygen levels led to the

EON	ERA	PERIOD OR SYSTEM	APPROX. AGE IN MILLIONS OF YEARS	
PHANEROZOIC	CENOZOIC	QUATERNARY	2	Age of Man
		TERTIARY	64	Age of Mammals
	MESOZOIC	CRETACEOUS	135	Age of Dinosaurs
		JURASSIC	200	
		TRIASSIC	230	
	PALEOZOIC	PERMIAN	290	Naked seed plants Earliest ancestors of dinosaurs
		PENNSYLVANIAN	300 320	First reptiles
		MISSISSIPPIAN	350	Ferns and seedless plants
		DEVONIAN	410	Age of Fish First Amphibians appear
		SILURIAN	440	First land plants Crabs, spiders, scorpions
		ORDOVICIAN	490	Corals & nautiloids appear
		CAMBRIAN	520 570	First fish Marine invertebrates Trilobites
PRECAMBRIAN	PROTEROZOIC	PRECAMBRIAN	670	First segmented worms & jellyfish
			2500	All fossils microscopic except for stromatolite mounds.
			3500	Warrawoona Microfossils
			3600	Oldest stromatolites Organic compounds (Origin of Life?)
			3800	Oldest sedimentary rocks
			4600	Origin of Earth

NOT TO SCALE

Geological time chart.

formation of deposits of iron oxide, or rust. These distinctive rocks are known as Banded Iron Formations. Today, these iron formations comprise the largest deposits of iron on the planet.

Two billion years after the formation of the Earth, cells and bacteria had become more specialized. Fossils found in rocks called the Gunflint Formation, 2 billion years old, bear a close resemblance to types of modern-day bacteria and blue-green algae.

By the end of the Precambrian, 600 million years ago, life had specialized and the seas teemed with life. A multitude of forms had evolved: jellyfish, flatworms, segmented worms. We can see their tracks preserved in the ancient rocks.
in the ancient rocks.

The fossil record became more complete as these animals secreted shells to protect their bodies from predators. The first appearance of animals with skeletons brings the Precambrian to a close and introduces the Paleozoic Era.

The Paleozoic Era

The *Paleozoic Era*, which lasted approximately 345 million years, is made up of the Cambrian, Ordovician, Silurian, Devonian, Mississippian, Pennsylvanian and Permian Periods.

Cambrian

The *Cambrian Period* lasts about 70 million years from 570 to 500 million years ago. The *Early Cambrian* is marked by the appearance of trilobites, a class of arthropods. (The horseshoe crab is a descendant of these ancient creatures.) The first molluscs appeared in the Cambrian. *Lingulella*, a tiny brachiopod, first appeared in the Cambrian. Related to the living Lingula, it is relatively unchanged after 550 million years. Unlike the Precambrian Era, in which

Lingulella

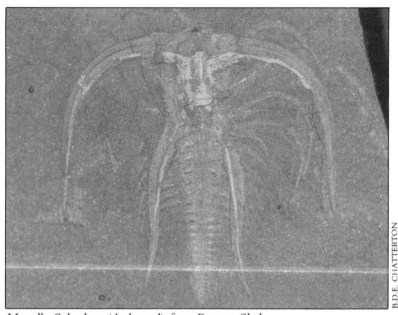

Marrella Splendens (Arthropod) from Burgess Shale.

most of the rocks are barren of fossils, the rocks deposited during the Cambrian are teeming with remains.

The Burgess Shale from Mt. Stephen near Field, British Columbia, contains a unique *fossil assemblage* in which soft worms and marine *invertebrates* (animals without backbones) are found preserved as fossils in fine-grained shales 550 million years old. This fossil find was the remains of a community living on the ocean floor which was rapidly buried by an undersea mudslide. Fine sediments covered them and prevented their being disturbed by scavengers. These fossils give us our best insights into life during the *Middle Cambrian*.

Ordovician

The *Ordovician Period* was 70 million years in duration, from 500 to 430 million years ago. It is marked by a greater diversity of life and it was during the Early Ordovician that the corals first evolved. Many earlier forms of trilobite became extinct. Other forms of life making an appearance were additional families of trilobites and nautiloids. While trilobites were mud feeders, the

Nautiloids

nautiloids (related to modern-day *Nautilus*, octopus and squid) had large brains and well-developed eyesight, and were specialized as predators.

Silurian

The *Silurian Period* is marked by the first abundance of fishes and extends from 430 to 395 million years. Crinoids (sea lilies) related to modern-day echinoderms such as starfish, and sea urchins appear here in the fossil record. The ancestors of crabs, lobsters, scorpions and spiders appear, and these forms all boast well-developed eyes and an increasing complexity over earlier types.

The Silurian produces the first record of land plants. The fossil remains of the first true land plant (*Cooksonia*, 5 centimetres) appear to be similar to the modern genus *Psilotum*. While *Cooksonia* probably lived in areas of standing water or tidal margins, this appears to be a step toward the successful colonization of the land.

The first animals to venture out of the sea onto the land were the arthropods. Primitive scorpions such as *Palaeophonus* are very similar to living species. Its jointed external skeleton, typical of all arthropods, was ideal for life on land. Fossils of mites, ticks

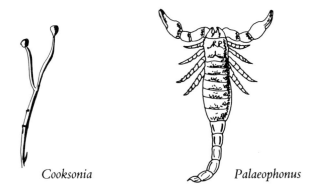

Cooksonia   *Palaeophonus*

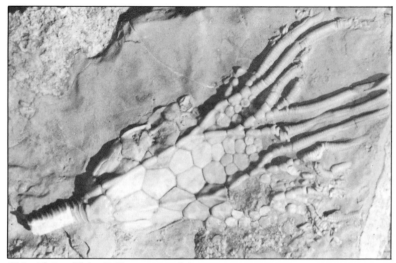
Crinoid (sea lily). Collection U of A.

and spiders have also been found in Silurian rocks which seem to indicate that there was an invasion onto the land to exploit available niches in the emerging plant cover.

The first fossils of fish related to modern forms are found in the Lower Silurian. The early fish, *Cephalaspis*, 18 centimetres long, is believed to have been a bottom dwelling mud feeder.

Devonian

The *Devonian Period* is known as the Age of Fishes. It lasted 50 million years, from 395 to 345 million years ago. During the Devonian a great burst in the evolution of the fishes took place. The evolution of lungs, which still survive in living fossils such as the African lungfish, and lobed fins found in the *Coelacanth* (discovered some years ago to be still living in the Indian Ocean), took place during the Devonian.

The development of lungs and limbs ideal for walking on land are seen in the first amphibian, *Ichthyostega*. It first appeared in the Upper Devonian and was about 1.5 meters in length. Adapted for terrestrial life, it had a flat skull, an adaptation to a life out of water, and short, powerful limbs which allowed it to crawl about on the land. Like all amphibians, it was tied to the water. Its weak link was the nature of its eggs. Unlike reptile and bird

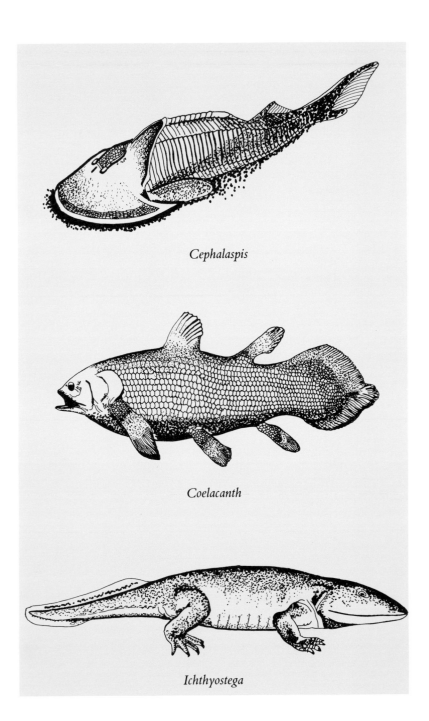

Cephalaspis

Coelacanth

Ichthyostega

eggs, which have a hard outer shell, amphibian eggs are little more than a sac of jelly surrounding the embryo. Most amphibian eggs need to be laid in water. Like modern-day salamanders and toads, most amphibians need to be kept moist. Therefore, it seems probable that early amphibians like *Ichthyostega* lived near the margins of freshwater.

There was a proliferation of land plants during the Devonian. Plants like ferns, horsetails, and club mosses all first made their appearance. Needing damp conditions like swamplands for successful reproduction, these forms still inhabit similar environments 350 million years later.

Mississippian and Pennsylvanian

During the *Mississippian* (345 to 325 million years) and the *Pennsylvanian* (325 to 280 million years) land plants evolved into huge forms and covered the land. The great coal-producing areas of the eastern United states resulted from the remains of these huge tropical forests being buried in the extensive coal swamps which flourished during the Pennsylvanian. Giant trees like *Lepidodentron* (30 metres) and *Calamites* (30 metres) grew in the swamps while giant dragonflies flew on 60 centimeter wings.

While *Ichthyostega* first appeared in the Devonian, it was during the Mississippian that changes in the environment saw a great diversity in the types and numbers of amphibians. Competition between types forced certain amphibians away from the damp

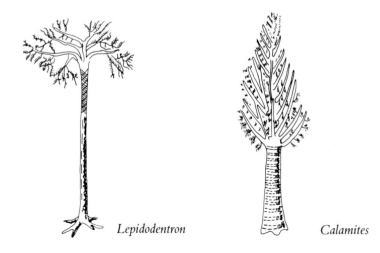

Lepidodentron   *Calamites*

Eryops, a prehistoric amphibian from the Early Permian, about 1.5 m in length. It may have lived both in the water and on land, similar to modern crocodiles (cast). Tyrrell Museum.

areas into the drier regions. This caused destruction of the amphibian eggs due to drying out. This selective pressure favoured the development of a self-contained egg with a strong membrane or outer shell which protected the developing embryo from drying out. The first animal to develop such an egg was the first reptile.

Permian

The *Permian Period*, 280 to 225 million years ago, saw the appearance of the *gymnosperm* (naked seed) plants and earliest ancestors of the dinosaurs. First appearing in the Pennsylvanian and Early Permian, early reptiles evolved into several lines which gave rise to different descendants. This classification of reptiles is grouped into four types: anapsids, synapsids, euryapsids, diapsids, and is based upon the shape and bone structure of their skulls.

The oldest known *anapsid* reptile (*Hylonomus*) is preserved as fossil remains found in Nova Scotia, Canada. This tiny reptile was about 20 centimetres in length and was found in fossilized

Hylonomus

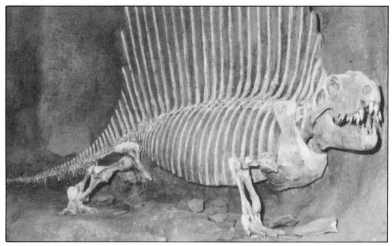

Dimetrodon, a large (3.5 m) flesh-eating reptile which lived from the Early to Middle Permian (cast). Tyrrell Museum.

tree trunks. The only living relative of this ancient anapsid line of reptiles are the turtles and tortoises.

Synapsid reptiles are thought to have been the ancestors of all mammals. This early type of synapsid reptile is represented by the fossil *Dimetrodon* (3.5 metres long), from the early to mid-Permian in Texas and Oklahoma.

Euryapsids are represented by reptiles like the extinct plesiosaurs. (See Marine Reptiles.)

Diapsid reptiles gave rise to two main subgroups. One subgroup consists of snakes and lizards (for example, the mosasaur; see Marine Reptiles), while the second subgroup includes the *thecodonts*, the ancestors of the dinosaurs. This subgroup is represented by *Petrolachosaurus*, the first known diapsid from the Permian rocks of Kansas in the United States. It is thought to have been a quick and agile predator, feeding on insects in the drier upland areas surrounding the coal swamps.

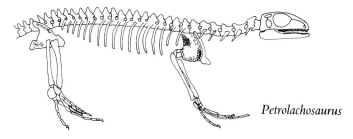

Petrolachosaurus

The Mesozoic Era

The end of the Permian brings the Paleozoic Era to an end. The Mesozoic Era or "Middle Age" was about to begin. This Era is made up of the *Triassic Period*, 225 to 190 million years ago, the *Jurassic Period*, 190 to 136 million years ago, and the *Cretaceous Period*, 136 million to 65 million years ago. This was the great era of the dinosaur. From common reptilian ancestors, the dinosaurs diverged into a powerful dynasty which would rule the earth for 160 million years. What follows is their story.

Dinosaurs found in Jurassic rocks
1. Allosaurus
2. Apatosaurus
3. Brachiosaurus
4. Camarosaurus
5. Diplodocus
6. Stegosaurus
7. Supersaurus
8. Ultrasaurus

Exposed Jurassic Rocks.

Note: Not all Jurassic rocks contain dinosaur fossils. Some of these outcrops were formed under the sea. This map is intended to serve as a general guide.

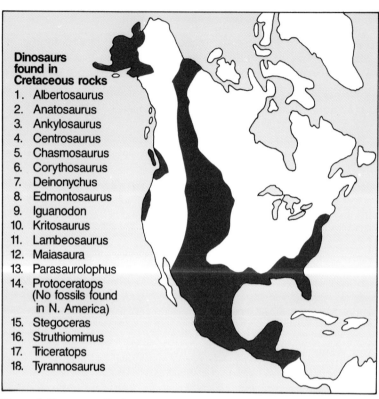

Exposed Cretaceous Rocks.

Note: Not all Cretaceous rocks contain dinosaur fossils. Some of these outcrops were formed under the sea. This map is intended to serve as a general guide.

PART II

The Bestiary

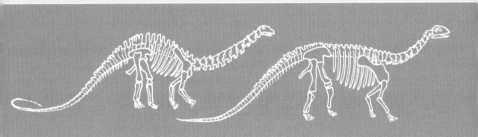

An Introduction to the Dinosaurs

The word "dinosaur" comes from the Greek word, *dinosauria*, which means "terrible lizard" (*deinos*, meaning terrible, and *sauros*, meaning lizard). The dinosaurs were related to a very wide range of reptiles, some of which are still living today, such as modern-day snakes, lizards and crocodiles. All reptiles have scales and cold blood. *Cold blood* means that unlike mammals, reptiles have no means of generating heat within their bodies. The temperature of a reptile is the same as its surroundings. If the surrounding air is warm the reptile is active; however, if it becomes too hot, the reptile will perish unless it can escape from the extreme heat. This is why most reptiles are inactive during

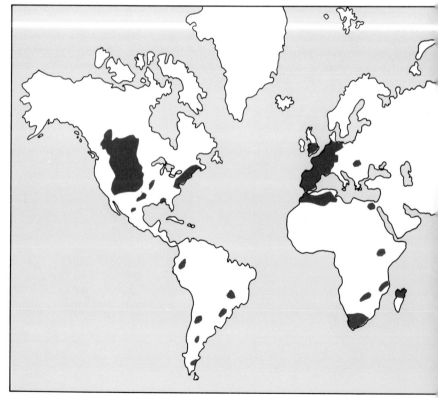

Significant dinosaur discoveries in the world.

the hottest part of the day. When it is cold, the reptile is very sluggish, or slow. This can be a distinct disadvantage in cold climates when compared with mammals, who can remain active at lower temperatures.

There were two orders of dinosaur: *Saurischia*, meaning lizard-hipped, and *Ornithischia*, meaning bird-hipped.

Saurischia

Saurischia consist of two suborders. The first of these are the *theropods* (beast feet) which include the flesh-eating dinosaurs. These dinosaurs were *bipeds* (meaning they walked on two feet) and ranged in size from lizard-catchers no larger than a chicken, to giants like *Tyrannosaurus rex* that weighed more than 6.5 tonnes.

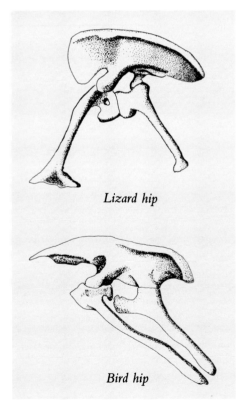

Lizard hip

Bird hip

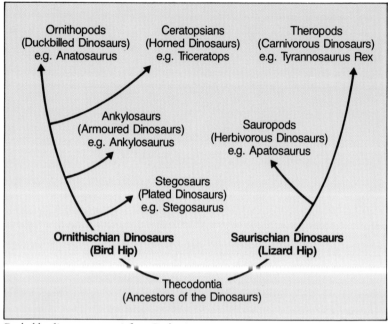

Probable dinosaur tree (after Corbett).

The hip bones of these dinosaurs are similar to those of modern-day lizards. They had strong hind limbs for running and smaller front legs or arms, suitable for holding prey.

The second suborder of Saurischia are the *sauropodomorphs* (lizard-feet forms), which include the largest land animals that ever lived. These giant creatures were *herbivores* or plant-eaters which walked on four legs. Like the theropods, the hip bones of sauropodomorphs resemble those of modern lizards. An example of this suborder is *Brachiosaurus* (23 metres).

Ornithischia

Ornithischia were mainly plant-eaters. The hip bones of these dinosaurs resemble those of modern-day birds. Like Saurischia, Ornithischia is grouped into two suborders. These are the two-legged *ornithopods* (bird feet), an example of which is *Stegoceras* (2.5 metres long); and the four-legged, horned, armoured and plated dinosaurs, an example of which is *Ankylosaurus* (10.7 metres long).

Albertosaurus (Alberta Lizard)

Saurischia: Theropoda
Late Cretaceous

Alberta, Canada
Montana, United States

Albertosaurus was a *carnosaur* (flesh-eater) and a relative of the larger *Tyrannosaurus rex*. *Albertosaurus* reached lengths of 8 metres and stood as high as 3.5 metres. The weight of a full-grown adult *Albertosaurus* could attain as much as 1.8 tonnes.

The general body shape of the *Albertosaurus* was similar to that of *Allosaurus* of the Late Jurassic. A *bipedal* (two-legged) dinosaur, *Albertosaurus* stood erect on two massive rear legs which ended in a three-toed foot, each carrying a sharp, curved talon. These razor sharp talons or claws were ideally suited for ripping, slashing or securing a hold on any potential prey animal. The tail, which accounted for one-half of the entire length of *Albertosaurus*, was carried off the ground when walking to provide a counterbalance for its massive body. Like earlier carnosaurs, the forelimbs of the *Albertosaurus* were very small and feeble and ended in a two-fingered hand. It seems unlikely

Albertosaurus skull. U of A Collection.

that these limbs could have been used to capture prey.

The probable lifestyle of *Albertosaurus* was like that of other carnosaurs. Most likely it was an opportunistic scavenger, driving off other smaller dinosaurs away from their kills, or feeding on the carcasses of dead animals. It is possible that *Albertosaurus* might have preyed on the common hadrosaurids (duck-billed dinosaurs), but, considering its size, it is unlikely it could have run down its prey on the open plain. Although its rear legs were very long and powerful, given the limitations of a reptilian metabolism (cold-blooded), it is more likely *Albertosaurus* captured its prey after short bursts of speed. Like modern reptiles, *Albertosaurus* may have been a lurking predator, lying prone to conserve energy, then, when the opportunity arose, it would rush from the shadows to capture its prey.

From the remains of juvenile *Albertosaurus* that have been found, it is possible that the slow-moving lifestyle of the adult might not hold true for young *Albertosaurus*. Given their light build, and the length of the legs in proportion to its body size, it is possible that the

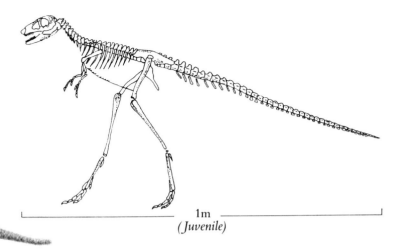

1m
(Juvenile)

Fossil carnosaur tooth impression.

juveniles might have captured small lizards, or smaller theropods, on the run. This would seem to indicate that the feeding habits of the youngsters would fill a different ecological niche than that of the adults.

Albertosaurus is the most common theropod fossil found in Alberta, and it must have been a very successful predator of the Upper Cretaceous, judging by the number of fossil remains found in the badlands of Dinosaur Provincial Park.

Allosaurus (Strange Reptile)

Saurischia: Theropoda
Late Jurassic

United States
Tanzania, Africa

The *Allosaurus* was a carnosaur which lived during the Late Jurassic. Reaching lengths of 12 metres and standing 4.9 metres tall, an adult *Allosaurus* may have weighed as much as 2 tonnes.

These dinosaurs were powerfuly-built bipeds that stood erect on massive rear-legs which ended in bird-like feet. The powerful leg muscles and razor-sharp talons were used to hold and dismember prey. The forearms of the *Allosaurus* were very small and ended in hands consisting of three fingers tipped with curved claws, it is unlikely these feeble limbs would have been useful for holding or catching prey. Some scientists think these small arms may have been useful in steadying the front quarters of the animal as it attempted to rise from a prone position. The tail of the *Allosaurus* was one-half of its body length and served to counterbalance the weight of the body as it stood erect.

The *Allosaurus* probably led a dual lifestyle much like modern-day large cats. If food was scarce, an *Allosaurus* might live as a solitary scavenger much like modern lions. Just as lions frequently chase other animals away from kills, it is very likely the *Allosaurus* would have also driven off other smaller theropods from carcasses. This practice of scavenging may have been seasonal.

The *Allosaurus*, like the Albertosaurus, may also have waited in ambush, relying on surprise to capture its prey. From what scientists surmise

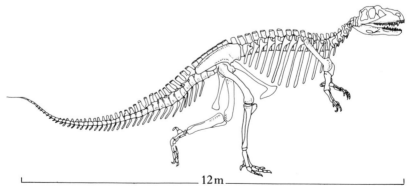

12 m

from modern reptiles, cold-blooded animals cannot maintain high energy output for long periods of time. Given the size of an adult *Allosaurus* (weighing 2 tonnes), paleontologists think it may have captured its prey in a short burst of speed from hiding, rather than a prolonged chase. It is unlikely that it could maintain this speed over a great distance. However, this might not be the case for juvenile *Allosaurus*. Considering their much smaller and lighter build, these young dinosaurs may have filled a different ecological niche than the larger adults, and may have captured their prey on the run.

The adult *Allosaurus* may have charged at its potential prey with its jaws agape, using its teeth to inflict mortal wounds, sending its victim into instant shock, thereby preventing its escape. The skull and massive neck muscles were perfectly adapted to absorb the force of such an impact. The skull of the *Allosaurus* was 60 to 90 centimetres long and its teeth were razor-sharp, serrated, and superbly adapted for tearing and cutting flesh. These sharp teeth were perfectly designed for attacking, killing, and dismembering the carcasses of prey animals. The shock of being stabbed with dozens of

Allosaurus (cast). Tyrrell Museum.

serrated daggers being propelled by 2 tonnes of *Allosaurus* should have been instantly fatal.

At other times, when prey was plentiful, *Allosaurus* may have banded together into packs to hunt large sauropods, such as *Apatosaurus* and *Diplodocus*. However, it is likely that an adult *Apatosaurus* (weighing 30 tonnes) would have been too large a victim for an attacking pack of *Allosaurus*, but juvenile sauropods would have been at great risk and were therefore the probable prey.

The Allosaurus died out at the end of the Jurassic Period. Many types of dinosaur died out at this time, and were replaced by the later dinosaurs which flourished during the Cretaceous. There are no clear answers to explain these extinctions (see Death of the Dinosaurs).

Anatosaurus (Duck Reptile)

Ornithischia: Ornithopoda
Late Cretaceous

Alberta, Canada
Wyoming, United States

The *Anatosaurus* was a very common hadrosaur during the Late Cretaceous in Alberta. A full-grown adult measured from 10 to 13 metres in length and weighed as much as 3 tonnes. The habits and lifestyle of *Anatosaurus* are thought to have been very similar to its close relative, *Edmontosaurus*. A herding animal, *Anatosaurus* may have fed upon conifer twigs, needles and seeds of the lowland forests. In the badlands of Wyoming, unusual fossil *Anatosaurus* mummies have been found. These are very rare fossils where both flesh and skin impressions have been preserved. Under examination, a fossilized stomach was found to contain seeds and conifer needles. From this rare find,  scientists are attempting to determine the diet of the hadrosaurs. This issue is still being studied.

The teeth of *Anatosaurus* are similar to those of other hadrosaurs. These teeth (several hundred in each jaw) are ideally suited for pulverizing and grinding tough plant material. *Anatosaurus*, like *Edmontosaurus*, had no crest. It is thought this dinosaur may have also communicated by means of loud bellows made from inflating the skin around its broad mouth parts.

Dinosaur skin impression.

Ankylosaurus (Fused Lizard)

Ornithischia: Ankylosauria
Late Cretaceous

Alberta, Canada
Montana, United States

The largest of the armoured dinosaurs, a full-grown *Ankylosaurus* could reach lengths of 10.7 metres and weigh from 2 to 3 tonnes. Built close to the ground, *Ankylosaurus* was a *quadruped* (walked on four legs) with heavy, short legs. The rear limbs ended in large toes, while the forelimbs were similar, only slightly shorter. To protect the legs from attack, they were tucked in under the body and were partially shielded by the body armour.

Ankylosaurus' main defence was a series of bony plates which covered the back, thereby protecting the body with a living armour. Along the sides the bony armour was modified into sharp spikes which protected the flanks from attack. The tail was equipped with a large, bony mass or

Ankylosaurus skull, Gobi Desert.

knob on the end which could be wielded like a large club. This was indeed a formidable weapon and provided defence against such predators as *Albertosaurus* or *Tyrannosaurus rex*. The large spikes could also serve as lethal daggers (with 2 to 3 tonnes of weight behind them). Due to their large size and massive body armour, it seems certain these living tanks were quite secure as they fed

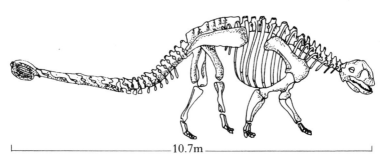

10.7m

upon low-lying vegetation like ferns and cycads.

Fossil skulls confirm that *Ankylosaurus* had a small brain with well-developed olfactory stalks which run to a well-developed nasal area. This may have aided the slow-moving dinosaur to detect danger.

Ankylosaurus died out during the mass extinctions of dinosaurs about 65 million years ago.

Apatosaurus (Deceptive Reptile)

Saurischia: Sauropoda
Late Jurassic

Utah, Oklahoma, Colorado, Wyoming, United States

Apatosaurus was a large sauropod dinosaur which lived during the later part of the Jurassic. Its remains are quite common in North America, first discovered in 1877 near Morrison, Colorado.

An adult *Apatosaurus* weighed approximately 30 tonnes, and was about 21 metres long from head to tail. The major portion of this length was comprised of its long neck and thin, tapering tail. There is a degree of uncertainty surrounding the skull of *Apatosaurus*. While *Apatosaurus* remains have been found, no skull has been recovered which was still attached. For years, skeletons of *Apatosaurus* had been restored using the *Camarasaurus*-like skull, but in recent years the issue was re-examined and it is now thought that the *Apatosaurus* skull resembled that of *Diplodocus*.

Apatosaurus is believed to have fed on the uppermost leaves of tall trees and low-lying vegetation growing on the margins of marshy areas. The dentition of *Apatosaurus* is composed of peg-like teeth which were used to cut vegetation before it was swallowed. The breakup and pulverizing of

this mass of vegetation was accomplished by stomach stones, or gastroliths. Fossils of *Apatosaurus* have been found with the stomach stones still present. These gastroliths were polished smooth by the constant action and churning of the muscular stomach.

The major predator of the *Apatosaurus* would have been *Allosaurus*. While it is unlikely an attack against an adult *Apatosaurus* would have been successful due to its large size, juvenile sauropods may have been at risk. It is believed that *Apatosaurus* travelled in herds for mutual protection and escaped to the safety of the water in case of attack. *Apatosaurus* became extinct during the mass extinctions at the end of the Jurassic period.

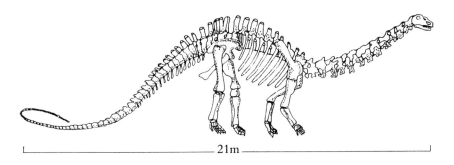

21m

Brachiosaurus (Arm Reptile)

Saurischia: Sauropoda
Late Jurassic

Colorado, United States
Tanzania, Africa

Brachiosaurus lived in North America and Africa during the Late Jurassic. One of the largest of the sauropod dinosaurs, a full-grown adult *Brachiosaurus* weighed as much as 102 tonnes and reached a length of as much as 23 metres. Like other sauropod dinosaurs, *Brachiosaurus* had an extremely long neck and tail, which accounted for approximately one-half of its total length.

The elongated front legs of *Brachiosaurus* were possibly an adaptation for browsing on the tall trees, much like modern day giraffes. The bones of the neck have spines which were likely used to secure elastic ligaments to raise the massive neck. This may have permitted *Brachiosaurus* to feed as high as 13 metres above the forest floor. Some scientists are of the opinion that *Brachiosaurus* raised itself up on its hind quarters to reach even higher into the canopy of the forest, using its powerful tail to counterbalance its huge body.

Questions about the diet of *Brachiosaurus* remain unanswered. Some theorize that *Brachiosaurus* fed on tall conifers. This mass of vegetation was stripped and passed along to the stomach where gastroliths pulverized the tough plant material. Bacteria may have

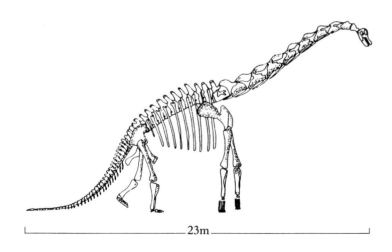

aided in this process.

Modern thinking in paleontology seems to support a terrestrial lifestyle rather than an aquatic existence for *Brachiosaurus*. In 1951, Professor Kenneth Kermack of London observed that the water pressure at 10 to 12 metres depth would have been sufficient to collapse *Brachiosaurus'* lungs; therefore ruling out deep water snorkeling.

It is possible that *Brachiosaurus* travelled in herds, gathering together for the protection of the young from such predators as *Allosaurus*.

Camarosaurus (Chambered Reptile)

Saurischia: Sauropoda
Late Jurassic.

Colorado, Oklahoma, Utah, Wyoming, United States

The *Camarosaurus* lived during the Late Jurassic. Remains have been discovered in Oklahoma, Wyoming, and at the Dinosaur National Monument in Utah.

Similar in general build to other sauropods, *Camarosaurus* is much smaller than *Brachiosaurus*. A full-grown adult *Camarosaurus* would have reached lengths of 18 metres, and weighed as much as 18 tonnes.

One feature of *Camarosaurus* is a short, blunt skull with extremely large nostrils positioned high on the head. At one time it was thought this feature indicated an aquatic lifestyle for *Camarosaurus* — living in deep water totally submerged with little more than its head breaking the surface. Today certain scientists speculate that

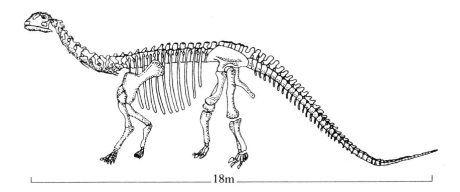

18m

Camarosaurus may have had a large trunk very much like modern-day elephants. This certainly would have made *Camarosaurus* a unique-looking dinosaur! Yet, perhaps a trunk is not as strange as one might think, had the sauropods lived on dry land rather than in marshes, a trunk would have been an asset for an animal which feeds in the trees. *Camarosaurus* could have used its large tail as a counterbalance to rise up on its hind legs and feed in the tree tops. This would allow a large animal to gather and consume huge amounts of foliage with little effort.

Its lifestyle was probably similar to that of other sauropods, herding together for defence of the young. Likely predators would have been *Allosaurus*; although a full-grown adult would have been at little risk from a single *Allosaurus*, however, young sauropod dinosaurs would have been very vulnerable to attack.

For reasons we do not understand, *Camarosaurus* declined in numbers, finally becoming extinct toward the end of the Jurassic.

Centrosaurus (Sharp Point Reptile)

Ornithischia: Ceratopsia
Late Cretaceous

Alberta, Canada

Fossils of *Centrosaurus* are found in the Red Deer River Valley in Dinosaur Provincial Park of Alberta. A common herbivore, a full-grown adult *Centrosaurus* would have reached lengths of 6 metres. Like other ceratopsids, *Centrosaurus* had a skull which flared toward the rear, covering the neck in a large, bony frill or shield. The top edge of the frill was covered with bony spurs, while the central portions of the shield are pierced with openings protected by two horns pointing downward from the top of the shield toward the snout.

Given its size, it is unlikely that *Centrosaurus* would have been able to flee from predators and may have gathered together in herds. In times of threat, they may have formed defensive circles, with larger animals forming the outer ring. This manner of defence protected the smaller animals and the young gathered toward the centre. While there seems to be some evidence for herding in some ceratopsids, it is not possible to say with any certainty the methods *Centrosaurus* used for defence.

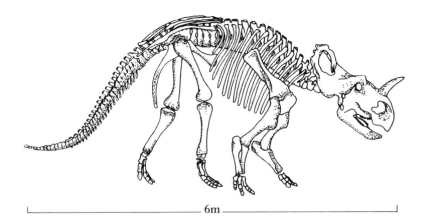

Centrosaurus skull. Collection U of A.

Chasmosaurus (Ravine Reptile)

Ornithischia: Ceratopsia
Late Cretaceous

Alberta, Canada
New Mexico, United States

Chasmosaurus was the earliest of the long-frilled ceratopsian dinosaurs. The skull bones of *Chasmosaurus* projected over the back, protecting the vulnerable neck with a hard, bony frill or shield. The frill was pierced by two large openings over the shoulder (probably filled with hard muscle during life). A second function of the shield was as an anchor to secure the powerful neck and jaw muscles. A herbivore, *Chasmosaurus* used its parrot-like beak to feed upon woody plant materials and ferns. The frill may also have functioned similar to antlers of modern-day moose or elk, that of sexual display and in courtship and mating rituals.

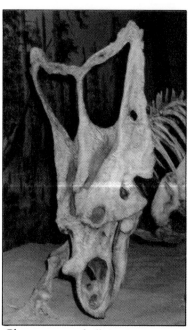

Chasmosaurus skull (cast). Tyrrell Museum.

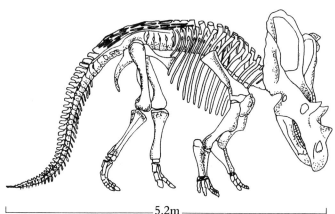

5.2m

Chasmosaurus was a solidly-built, slow-moving creature. A quadruped, *Chasmosaurus'* limbs end in four toes. Unlike the specialized hands of *Albertosaurus* or some of the hadrosaurs, the forelimbs of *Chasmosaurus* served merely to support its front quarters and were of little use in feeding.

Chasmosaurus was 5.2 metres long, somewhat smaller than its immediate relatives. Fossils of *Chasmosaurus* have been found in Dinosaur Provincial Park in Alberta.

Corythosaurus (Helmet Reptile)

Ornithischia: Ornithopoda
Late Cretaceous

Alberta, Canada

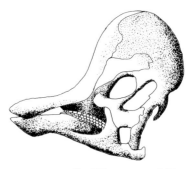

Corythosaurus was very common in Alberta during the Late Cretaceous. Its fossils have been found throughout the badlands. A rather large animal, the full-grown adult reached lengths of 10 metres and weighed as much as 3.8 tonnes. *Corythosaurus* was a crested hadrosaur, possessing elaborate enlargements of the skull. These chambers were connected to the nostrils of the hadrosaurs. The exact purpose of these outgrowths is not clear. Perhaps they evolved as a sexual characteristic (analogous to antlers in modern-day deer and elk) or served as resonating chambers for calling between members of a herd during courtship, similar in function the bellows of modern-day moose or elk. The crest of *Corythosaurus* resembled half a dinner plate stuck on top of its head.

Like other hadrosaurs, *Corythosaurus* may have gathered together in herds to protect the young from predators like *Albertosaurus*.

A herbivore, *Corythosaurus* probably fed upon conifer twigs, needles, ferns and cycads

Hadrosaur teeth. Note how the teeth are arranged in batteries.

growing along the marshes or throughout the woody lowlands. *Corythosaurus* was superbly equipped for shredding this tough vegetation, boasting hundreds of teeth arranged in batteries like files or rasps.

Corythosaurus was extremely abundant throughout Alberta during the Late Cretaceous, but declined in numbers, finally becoming extinct toward the end of the period.

Deinonychus (Terrible Claw)

Saurischia: Deinonychosauria
Late Cretaceous

Western United States

Deinonychus was discovered in Montana in 1964 by Professor John Ostrom of Yale University. This find of several skeletons provided a great understanding of this unique dinosaur. *Deinonychus* was an agile, bipedal dinosaur from 2.4 to 4 metres long.

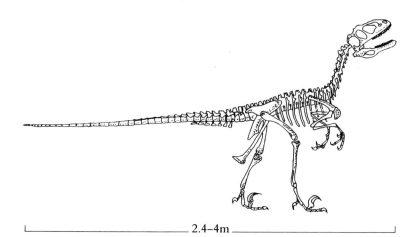

2.4–4m

Deinonychus likely walked erect using its tail as a counter-balance. Its hind legs each ended in three clawed toes, one of which was extra large and sickle-shaped, reaching lengths of 12.7 centimetres. While scientists cannot be certain, this large claw may have served as a slashing weapon. The forelimbs were powerful, long and flexible, ideal for holding prey.

Deinonychus also possessed powerful jaws, lined with backward-curving teeth.

Deinonychus, much like modern-day lions, may have hunted in packs, attacking young, injured or old sauropods. Other potential prey for these hunting packs would have been the *Ankylosaurus* or ornithopods like the hadrosaurids.

Diplodocus (Double Beam)

Saurischia: Sauropoda
Late Jurassic

Colorado, Montana, Utah, Wyoming, United States

Diplodocus was a large sauropod dinosaur living in North America during the Late Jurassic period. A close relative of *Apatosaurus*, it reached lengths of nearly 27 metres. Although longer than its close relative, a full-grown adult would have only reached weights of 10 to 11 tonnes.

The skeleton of *Diplodocus* was of lighter build than *Apatosaurus*. Its light skeleton resembled a cantilever bridge, with most of the weight borne by the massive bones of the shoulder. The pelvis, which was incredibly strong, was supported by five vertebrae fused together for added strength. The nostril openings were positioned on the top of the skull, while the teeth were thin and pencil-like, unlike the peg-like teeth of *Apatosaurus*.

It is thought that *Diplodocus* had similar feeding patterns to other sauropods, feeding upon low-lying vegetation bordering lakes and marshes and on the uppermost vegetation of tall trees. Unlike the *Apatosaurus*, stomach stones have not been

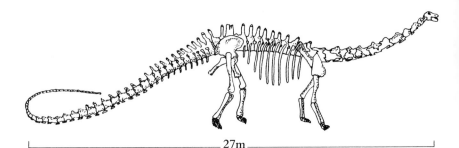

27m

found with *Diplodocus* remains. Perhaps these fossils were lost before deposition of the body in sediments. While it is not possible to speculate with any certainty on the mechanics of *Diplodocus'* digestion, it seems highly probable it was similar to *Apatosaurus* and other sauropods.

Like other sauropods, *Diplodocus* may have been a herding animal and in times of attack retreated to the water for protection from predators such as *Allosaurus*. Only juveniles would have been at risk; it is unlikely that adult *Diplodocus* were vulnerable to attack — its long tail, used as a giant whip, was a most effective weapon against any attacking predator.

The *Diplodocus* fell victim to the mass extinctions at the end of the Jurassic.

Edmontosaurus (Edmonton Reptile)

Ornithischia: Ornithopoda
Late Cretaceous

Alberta, Canada
Montana, United States

Edmontosaurus was a hadrosaur, an extremely successful group of ornithopods which evolved in the Late Cretaceous, and thrived right up until the great dinosaur extinctions at the end of the Cretaceous, 65 million years ago.

Edmontosaurus was quite common in Alberta and Montana during the Late Cretaceous. A full-grown adult could reach a length of 13 metres and a weight of 3.1 tonnes. The powerful rear legs of *Edmontosaurus* ended in three toes, allowing it to move upright on two legs. The forelimbs were shorter and may have assisted in feeding. *Edmontosaurus* is thought to have been a herding animal with groups gathering together to afford protection.

The herbivorous *Edmontosaurus* fed on tough plant material. Originally, paleontologists thought that the mouth parts, which were

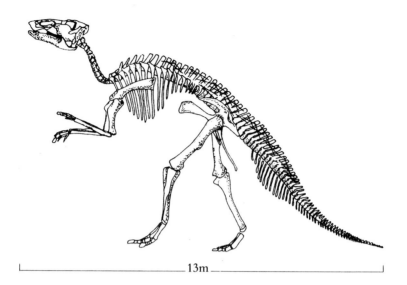

|—————— 13m ——————|

shaped somewhat like the bill of a duck, were an adaptation for feeding upon water plants and straining soft mud in swamps for tiny animals. The current thinking, based upon fossil evidence found in the fossilized stomach of a single hadrosaur, seems to indicate that the diet of *Edmontosaurus* and other hadrosaurs consisted of conifer needles, twigs and seeds of land plants. The teeth of the *Edmontosaurus*, like other hadrosaurs, consist of hundreds of interlocking teeth arranged in batteries much like a file or rasp. These were perfectly suited to shredding the tough land plants on which *Edmontosaurus* is thought to have fed.

Unlike some other hadrosaurs, *Edmontosaurus* did not have a crest. Rather, the area over the nostrils on the front of the skull may have been covered with loose skin which if inflated, may have produced a loud bellowing call similar to modern-day moose or elk. While scientists are not sure about this interpretation, these calls might possibly have served as mating calls, or as a means for a herd of *Edmontosaurus* to communicate.

Iguanodon (Iguana Tooth)

Ornithischia: Ornithopoda
Lower Cretaceous

Western Europe, Romania
Western North America
North Africa; Mongolia

Iguanodon was an extremely common dinosaur during the Early Cretaceous. Remains have been found extensively in Western Europe and Northern Africa.

A full-grown adult *Iguanodon* could reach lengths of 9 metres, stand 5 metres high, and attain a weight of 4.5 tonnes. *Iguanodon* had two massive hind legs, each bearing three toes, and probably walked erect using its powerful tail as a counterbalance to its massive body. One unique feature of *Iguanodon* was a specialized hand with four digits and a large vicious spike which may have been used to strip vegetation or as a defensive weapon against predators.

Fossil evidence suggests that *Iguanodon* was a herd animal. Accumulations of thirty skeletons known from a Belgian coal mine and tracks found in Germany and England lend support to the animal's

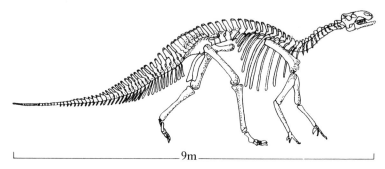

social behaviour.

The *Iguanodon* seems to have filled a niche not unlike that of the hadrosaurids (duck-billed dinosaurs) which evolved in the later Cretaceous.

Kritosaurus (Chosen Reptile)

Ornithischia: Ornithopoda
Late Cretaceous

North America

A full-grown *Kritosaurus* could grow to 9 metres long. Like other members of the hadrosaur family, *Kritosaurus* was a herbivore, likely feeding on lowland plants such as conifers, cycads and ferns which grew in the swampy lowland plains. Walking erect on its hind legs using its tail to counterbalance its body weight, the *Kritosaurus* had forelimbs which possessed fingers or digits, which may have been used in feeding. Like other hadrosaurs, it had batteries of interlocking teeth, useful for shredding or grinding vegetation.

The lifestyle of *Kritosaurus* was probably similar to other hadrosaurs, gathering together into herds for mutual protection of the young from predators such as *Albertosaurus*.

Fossil remains indicate that *Kritosaurus* became extinct during the latter part of the Cretaceous.

Lambeosaurus (Lambe's Reptile)

Ornithischia: Ornithopoda
Late Cretaceous

Alberta, Canada
Montana, United States

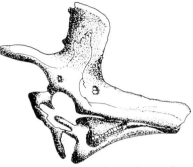

Lambeosaurus, named after a Canadian paleontologist, Lawrence Lambe, was one of the largest of the crested hadrosaurs. Specimens 15 metres long have been found. The large, square-shaped crest of *Lambeosaurus* was in some cases larger than the entire skull. Similar in build to other hadrosaurs, *Lambeosaurus* walked erect on two powerful hind legs and may have used its forelimbs for feeding. Living in forested lowlands, *Lambeosaurus* may have fed upon conifers, ferns and cycads. Like other hadrosaurs, it is thought that *Lambeosaurus* gathered together into herds to protect the young. The main predator was *Albertosaurus*.

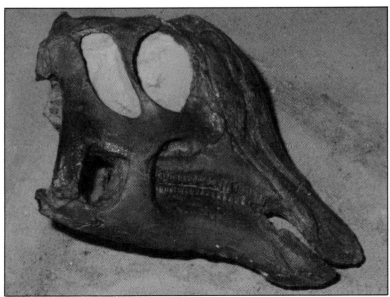

Juvenile Lambeosaurus skull (cast). Tyrrell Museum.

Maiasaura (Good Mother Reptile)

Ornithischia: Ornithopoda
Late Cretaceous

Montana, United States

Maiasaura was one of the hadrosaurs or duck-billed dinosaurs which were very common during the Late Cretaceous in western North America. Similar in build to other hadrosaurs, *Maiasaura* was about 9 metres long and walked erect on two hind legs. The forelimbs, like those of other hadrosaurs, were equipped with digits or fingers.

It seems likely that *Maiasaura* was a herbivore, feeding on woody plants, conifers, cycads and ferns in the woody lowlands. Dentition consisted of batteries of interlocked teeth used to shred vegetation. *Maiasaura* may have gathered together for protection of the young from predators like *Albertosaurus*.

Extraordinary finds of *Maiasaura* nests, eggs, juveniles and adults were discovered in Montana in 1978 and 1979. The eggs were arranged in circles and are slightly elongated in length, like a short sausage. The nest was approximately 2 metres wide and about .75 metres deep. Several nests were found together along with many shell fragments, possibly indicating a nesting colony was in use for many years. The evidence seems to suggest that *Maiasaura*, and possibly other dinosaurs, actively cared for their young, unlike most modern-day reptiles.

While a full-grown adult *Maiasaura* was 9 metres long, the fossil newly-hatched baby

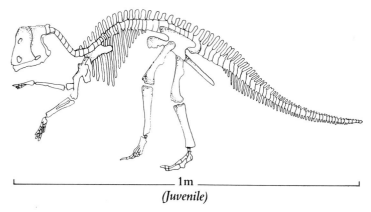
(Juvenile)

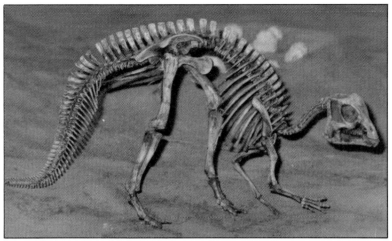
Skeletal restoration of infant Maiasaura (cast). Tyrrell Museum.

dinosaurs were only 50 centimetres in length. Larger juveniles (1 metre long) were also found in the vicinity of the nest. Judging by the wear of their teeth, and the fact that they are very much larger than the newly-hatched young, it is thought these larger young may have been cared for by adults as they foraged near the nests (see Dinosaur Eggs and Young).

Maiasaura became extinct toward the end of the Cretaceous.

Parasaurolophus (Reptile with Parallel Side Crest)

Ornithischia: Ornithopoda
Late Cretaceous

Alberta, Canada
New Mexico, United States

Of all the crested hadrosaurs, *Parasaurolophus* was the most striking. A full-grown adult attained lengths of up to 10 metres. The head crest extended back for a distance of up to 1.8 metres, and very much resembled a large snorkle perched atop its skull.

Parasaurolophus walked erect on two hind legs, each of which sported three large toes. The shorter forelimbs each possessed four digits useful for feeding. Like other hadrosaurs, *Parasaurolophus* was likely a herbivore feeding on lowland plants and shrubs such as ferns, cycads, conifers and other land plants. The dentition of *Parasaurolophus* consisted of hundreds of teeth arranged in batteries like files or rasps and used to shred bulky vegetation. Patterns of wear on fossil teeth suggest these animals lived on land rather than in swamps. Animals feeding on land vegetation have a tendency to ingest more sand than those living in silty or muddy terrain. This abrasive grit results in greater wear on the teeth of land dwellers.

Parasaurolophus possibly congregated into groups for the protection and rearing of the young.

At one time the belief was held that the crest of *Parasaurolophus* served as a snorkle when the animal fed along the bottom of marshes or swamps.

However, this theory has been discounted, as there was no breathing hole at the top of the crest. As with other crested hadrosaurs, the crest may have served during courtship rituals or for calling between members of the herd. Air moving through passageways in the crests would resonate, allowing for distinctive calls which were probably unique to each species.

Parasaurolophus died out during the dinosaur extinctions at the end of the Cretaceous, 65 million years ago.

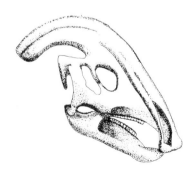

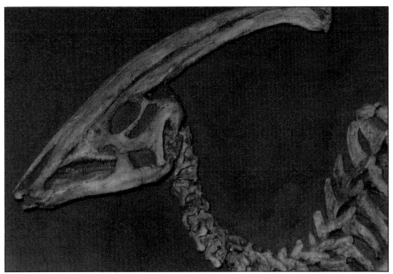

Parasaurolophus skull (cast). Tyrrell Museum.

Protoceratops (First Horned Face)

Ornithischia: Ceratopsia Mongolia, Asia
Late Cretaceous.

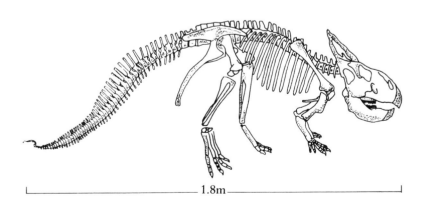

When did the *Protoceratops* live? It is difficult to state with any certainty. The beds in which its fossil remains have been found have not been firmly dated.

Fossils have been found in Late Cretaceous rocks of central Asia (Mongolia). The first *Protoceratops* was discovered during expeditions to the Gobi Desert in the 1920s. These finds were sensational for the individual specimens ranged in age from newly-hatched young to the very old. The most significant finds were the extremely well-preserved nests of dinosaur eggs. These were the first dinosaur nests ever discovered and were found in close proximity to the fossil remains of both adult and juvenile *Protoceratops*. This abundance of dinosaurs supports the theory of herding around the nesting site, similar to hadrosaur finds in Alberta and Montana. (See Dinosaur Eggs and Young.)

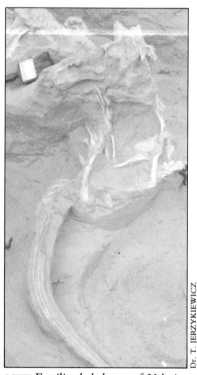 

LEFT *Fossilized skeletons of Velociraptor and Protoceratops, discovered in 1971 by the Polish–Mongolian expedition in the Gobi Desert at Toogreeg. This extremely rare fossil find documents a fight to the death.* RIGHT *Protoceratops.*

Protoceratops was a small, quadrupedal (walked on four legs) dinosaur approximately 1.8 metres long, weighing about 100 kilograms. The hornless frill or crest of *Protoceratops* likely evolved to protect the neck from attack, and to anchor the powerful jaw muscles.

A herbivorous dinosaur, *Protoceratops* used its parrot-like beak to shred plant material into bite-sized bits. The teeth, like those of hadrosaurs, were arranged in batteries. However, unlike the grinding action of the hadrosaur batteries, the teeth of *Protoceratops* formed vertical shearing blades which cut the food in an action similar to that of scissors. Did the *Protoceratops* use gastroliths to grind up its food? Paleontologists do not know for certain, as gastroliths have not been found in association with the fossil remains of *Protoceratops*.

Stegoceras (Horny Roof)

Ornithischia: Pachycephalosauria
Late Cretaceous

Alberta, Canada

Stegoceras is one member of a family of thick-headed dinosaurs which lived during the Late Cretaceous.

A small biped (two legs), *Stegoceras* walked erect on its two hind limbs. The tail was long, comprising one-half the total body length, and served to counterbalance the weight of its body. A lightly-built creature, *Stegoceras* was likely quick and agile, allowing it to escape from predators such as juvenile *Albertosaurus* or other small theropods.

A most unusual feature of these small dinosaurs (2.5 metres in length and weighing up to 54.4 kilograms) is the top of the skull which is thickened and forms a small dome. Some scientists have suggested this dome played a part in mating. If *Stegoceras* males charged each

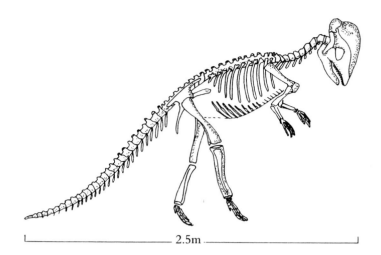

2.5m

other (like modern-day mountain sheep) to determine which would mate with females, then the thickened skull bones would have served a purpose in protecting the heads of combatants.

The *Stegoceras'* lifestyle may have been similar to that of sheep or goats: living in herds. However, this cannot be stated with any certainty, because the post-cranial skeleton of this dinosaur is known only by the fossil remains of one individual, found in the Red Deer River Valley in Dinosaur Provincial Park, Alberta, and currently in the collection of the University of Alberta. Other fossil finds include scattered skull fragments, usually the hard bony dome.

Stegoceras became extinct during the later part of the Cretaceous.

Stegosaurus (Plated Reptile)

Ornithischia: Stegosauria
Late Jurassic

Colorado, Oklahoma, Utah, Wyoming, United States

Stegosaurus is noted for its large, angular plates which stick up along its spine. A full-grown adult could reach lengths of up to 9 metres and weigh as much as 1.8 tonnes. The tail of *Stegosaurus* sported four bony spikes which may have served as its main defence against predators. Walking on four legs, *Stegosaurus* is thought to have been slow-moving and certainly not quick-witted — its brain was about the size of a walnut. Because the front legs of *Stegosaurus* are shorter than its rear legs, some paleontologists believe that it might have risen up on its rear legs (like modern-day elephants) and fed on vegetation growing in the tree tops.

The teeth of *Stegosaurus* were quite feeble and useful only for cutting vegetation rather than shredding or grinding it. Typical diet would have consisted of ferns, cycads and low-growing plants, or perhaps conifers, if there is any truth to the theory the animal rose up on its hind legs.

The plates which rise up from the spine may have doubled as armour or as heat shields to collect solar heat if *Stegosaurus* was cold-blooded. The bones of these plates were filled with many blood vessels which would have acted as a large radiator to warm the blood. Or, if in fact *Stegosaurus* was warm-blooded, then perhaps these plates served to cool the blood, much like the ears of modern-day elephants.

Stegosaurus became extinct toward the end of the Jurassic period.

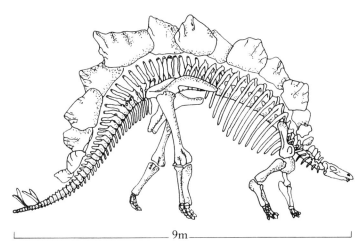

Struthiomimus (Ostrich Mimic)

Saurischia: Ornithomimosauria
Late Cretaceous

Alberta, Canada
New Jersey, United States

Struthiomimus very much resembled a modern-day ostrich without feathers. Small and agile, *Struthiomimus* (3.5 metres long, weighing 100 kilograms) walked erect on two hind legs; the forelimbs were long and slender and ended in three flexible fingers. Possessing large eyes and brain, *Struthiomimus* is thought to have been quick and agile, using these abilities to capture small reptiles, mammals and insects or to escape from predators.

Lacking teeth, *Struthiomimus* probably swallowed its food whole or in large chunks, using a gizzard with gastroliths to grind its food. However, this cannot be proven, as no known *Struthiomimus* fossil evidence for gastroliths exists. Like other members of this family, the diet of *Struthiomimus* was probably supplemented with eggs if the opportunity presented.

Very rare fossil remains of *Struthiomimus* were found near Steveville on the Red Deer River in Alberta, Canada in 1926. This

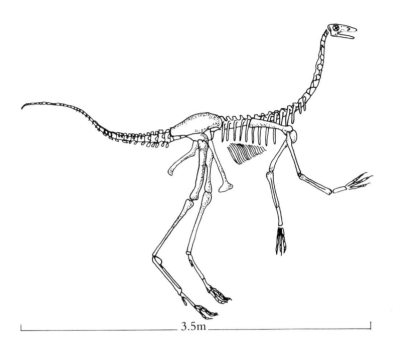

3.5m

was an unusual and significant find in that the skull was intact and a large portion of the skeleton was complete.

Struthiomimus survived up to the end of the Cretaceous and died out about 65 million years ago during the great extinctions which closed the Age of the Dinosaur.

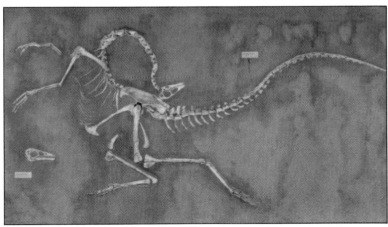

Struthiomimus (cast). Tyrrell Museum.

Supersaurus (Super Reptile)

Saurischia: Sauropoda
Late Jurassic

Colorado, United States

Discovered as recently as 1971, the bones of *Supersaurus* seem to indicate that it may have resembled *Brachiosaurus*, though longer, and lighter in weight. *Supersaurus* may have been as long as 30 metres long and as high as 16.5 metres. The length of neck has been estimated to be 12 metres, while the largest vertebra was over 1.5 metres long. The fossilized shoulder blade (2.4 metres) was longer than a human.

Little is known about *Supersaurus*, but it is thought that it was similar to other sauropods. *Supersaurus* has yet to be named scientifically, and is still being studied.

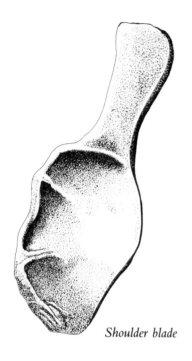

Shoulder blade

Triceratops (Three Horned Face)

Ornithischia: Ceratopsia
Late Cretaceous

Western Canada
United States

One of the best-known members of the family of horned dinosaurs, *Triceratops* was also the largest member of the ceratopsids. A full-grown adult *Triceratops* may have weighed as much as 5.4 tonnes and reached lengths of up to 9 metres. A quadruped, *Triceratops* moved slowly on four massive legs, each ending in four toes, similar in construction to those of elephants or rhinos. Lacking any specialized digits for grasping vegetation, the forelimbs would have been of no use in feeding. The skeleton of *Triceratops* is solidly-built to support its great weight. The first three neck vertebrae were fused together to support the weight of the massive skull.

Triceratops (cast). Tyrrell Museum.

The bony shield of *Triceratops*, unlike *Centrosaurus*, is not pierced by holes but is solid bone, with two very large horns just over the eyes and a smaller third horn on the snout. Horn cores (the bone centre) have been found as long as 90 centimetres, indicating the horns would have been much longer during life due to a horn sheath which fitted over the bone core. These horns would have been formidable weapons. Like other ceratopsids, *Triceratops* likely stood and fought rather than flee from predators. A 5-tonne *Triceratops* charging head-down would have been more than a match for any predator. Even *Tyrannosaurus rex* and *Albertosaurus* likely preyed upon the young or sick and dying members of a herd, rather than full-grown, healthy adults.

While it is not possible to determine certain characteristics of dinosaur behaviour based solely on fossil evidence, some scientists are of the opinion that herds of *Triceratops* may have formed defensive circles protecting the young. While this cannot be proven, no doubt a ring of metre-long horns would have stopped the most determined predator.

Triceratops used its powerful parrot-like beak to tear and shred plant material, while its powerful jaw muscles (attached to the great frill) provided the strength to chop up the toughest plants. Gastroliths aided digestion.

Very successful and with few enemies, *Triceratops* survived right up until the end of the Age of Dinosaurs, and was one of the last to become extinct 65 million years ago.

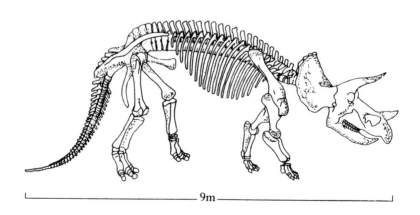

9m

Tyrannosaurus (Tyrant Lizard)

Saurischia: Theropoda
Late Cretaceous

Alberta, Canada
Montana, United States
China

Tyrannosaurus rex lived during the Late Cretaceous, and was the largest carnivorous land animal that ever walked upon the earth. Standing 6 metres high, this mighty creature reached lengths of up to 15 metres and weighed an astonishing 6.5 tonnes.

Tyrannosaurus rex fossils are very rare. Fewer than ten skeletons have ever been found. Carnivorous dinosaurs comprised no more than 5 percent of the total animal population, and *Tyrannosaurus rex* was only one type.

Tyrannosaurus rex was bipedal and stood erect on two massive rear legs which ended in three-toed feet, each toe possessing a single hooked claw or talon. The foot of a modern-day turkey, though much smaller, is similar in design to the foot of *Tyrannosaurus rex*. The animal's massive body was counterbalanced by its huge tail, which was carried erect as the animal walked. Like other tyrannosaurids, the forelimbs of the *Tyrannosaurus rex* were small and feeble, each ending in a two-fingered hand. It is

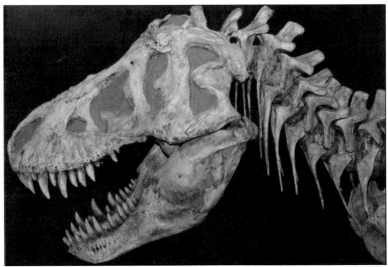

Tyrannosaurus skull (cast). Tyrrell Museum.

unlikely these small forearms were of use in holding or capturing prey.

The powerful skull (over 1.2 metres long) was of light construction and pierced with holes, thus reducing the overall weight. The jaw muscles were

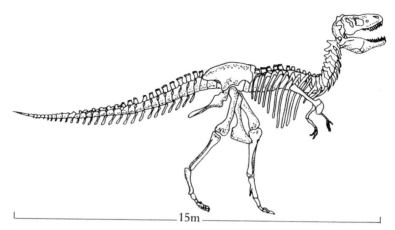

likely massive. The teeth were serrated and up to 18.4 centimetres long. These teeth were perfectly adapted for a scavenger feeding on dead carcasses.

Ultrasaurus (Ultra Lizard)

Saurischia: Sauropoda
Late Jurassic

Colorado, United States

Discovered in 1979 near the fossil site of *Supersaurus*, *Ultrasaurus* may have been the largest dinosaur which ever lived. Fossil evidence suggests a beast of up to 30.5 metres long with a weight of as much as 136 tonnes. Little is known about this dinosaur; however, its patterns of behaviour were likely similar to that of other large sauropods. *Ultrasaurus* has not yet been scientifically named and is still being studied.

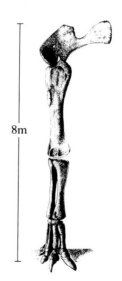

Archaeopteryx (Ancient Wing)

Late Jurassic

In 1861, the fossilized remains of a small, birdlike creature about 1 metre long were discovered in a limestone quarry near Solnhofen, Germany, by Dr. Karl Haberlein. The fossil was acquired by the British Museum (Natural History). What is incredible about the fossilized remains of *Archaeopteryx* is the impression of feathers preserved in the fine-grained lithographic limestone. Shrouded in controversy since its discovery, *Archaeopteryx* seems to lend support to the theory that birds and reptiles are closely related.

The Solnhofen find reveals that *Archaeopteryx* had socketed teeth in the jaws, a long, bony tail, and claws on its feathered forelimbs — characteristics suggesting a reptile ancestry. And yet, like modern birds, *Archaeopteryx* had long forelimbs proportioned like wings, a wishbone in its breast, and very large eyes and brain. Birds, like reptiles, have scales on their legs, and both birds and reptiles lay shelled eggs.

Birds may have evolved from primitive *arboreal* reptiles, whose feathers developed as a mechanism for breaking falls, or allowed them to glide from tree to tree. The insulating properties of feathers, taken in light of modern theories about warm-blooded dinosaurs, suggest that perhaps, like small

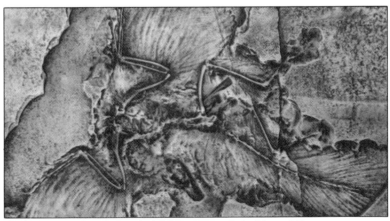

Archaeopteryx (cast). Provincial Museum and Archives, Edmonton.

theropod dinosaurs, *Archaeopteryx* was warm-blooded. Only six specimens of *Archaeopteryx* have ever been found, one of which resides in the British Museum in London. A second is in the Museum of Natural History in East Berlin. A small third specimen, discovered in 1951 and originally labelled *Compsognathus*, has no feather impressions. Not until 1973 was the specimen confirmed as being an *Archaeopteryx* by Dr. Peter Wellnhofer.

Best guesses regarding *Archaeopteryx* suggest it was probably a poor flyer, and merely used its wings for short, gliding hops, or to assist it in trapping small insects or reptiles, somewhat in the manner of modern bats. Some paleontologists suggest that perhaps *Archaeopteryx* was merely a feathered dinosaur, and not a direct ancestor of birds.

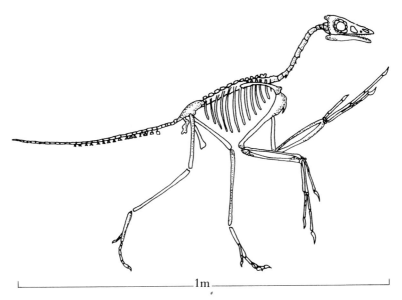

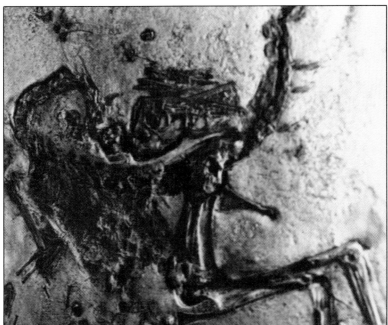

Compsognathus (pretty jaw) was similar in build to Archaeopteryx. This tiny dinosaur was about as large as a chicken and weighed about 3 kg (6.5 lbs.). (Cast). Tyrrell Museum.

Pterosaurs (Winged Reptiles)

Pterosaurs are not true dinosaurs, but rather "winged reptiles". The earliest examples of pterosaur originated in the Late Triassic about 200 million years ago. *Dimorphodon*, an early Jurassic pterosaur about 1.5 metres long, sported a long tail, a large head and jaws equipped with well-developed teeth. The diet of *Dimorphodon* was most likely insects or small vertebrates which it caught on the wing.

Paleontologists are of the opinion the small pterosaurs may have been active flyers whereas the larger species were most likely gliders. The wings had three clawed fingers about the midpoint of the wing, while a fourth finger was extremely long and formed the support for the rest of the wing. Pterosaurs possessed a breastbone similar to modern birds. This provided the attachments for muscles to power their flight. The rear legs ended in clawed feet. Perhaps pterosaurs were able to walk about by folding their leathery, membranous wings clear of the ground. This theory is controversial, and not widely accepted.

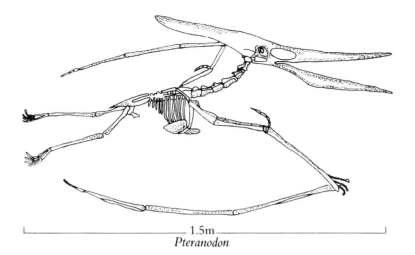

Pteranodon

Dimorphodon

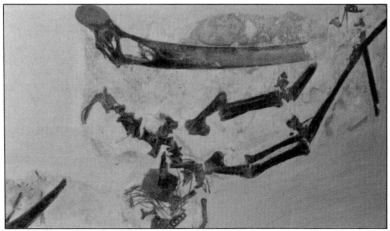

Fossil of Pterosaur (Pteranodon). Collection U of A.

The largest pterosaurs evolved during the Late Cretaceous. *Pteranodon* (winged and toothless) had a wingspan up to 7 metres, and was considered the largest of the pterosaurs. However, this idea was to change with the discovery of ultra-large pterosaurs at Big Bend National Park in Texas.

Discovered by Douglas Lawson, these remains were named for the Aztec god, Quetzalcoatlus or "feathered serpent" has an estimated wingspan of 15 metres.

These fascinating reptiles, for all their strange beauty and flying abilities, became extinct at the end of the Cretaceous about 65 million years ago.

Marine Reptiles

Reptiles known as ichthyosaurs, plesiosaurs and mosasaurs evolved during the Mesozoic Period. While not dinosaurs, these important reptiles deserve examination.

Ichthyosaurs (fish reptiles) first appear in the fossil record during the Early Triassic about 200 million years ago. The body shape of ichthyosaurs is similar to dolphins or killer whales. The jaws are long and pointed, and lined with sharp

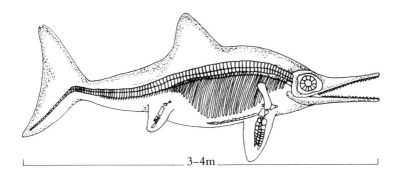

teeth. Ideally suited for sea life, it is thought ichthyosaurs fed on fish, cephalopods, and even on plesiosaurs.

Unlike reptiles on land which laid eggs, ichthyosaurs gave birth to live young. Fossils have been found with embryos in the body cavities of female ichthyosaurs. The most amazing of these finds is that of a female ichthyosaur with a baby partially born. Half of its body is still within its mother; perhaps complications at birth caused the death of the mother.

Plesiosaurs (ribbon reptiles) were less fishlike than the ichthyosaurs and some species had very long necks and large feet, shaped like paddles, ideally suited for life in the water. The jaws of plesiosaurs were lined with sharp teeth and most paleontologists are of the opinion that plesiosaurs fed upon a diet of fish, other plesiosaurs, turtles and ammonites (related to the *Nautilus*).

Neither ammonites nor plesiosaurs survived the extinctions at the end of the Cretaceous.

Mosasaurs (reptiles from Meuse) are very common in Cretaceous rocks. Feeding on fish and shellfish, these marine lizards were very large, reaching lengths of 9 metres. The jaws were lined with sharp teeth and the feet evolved into large paddles, ideal for an

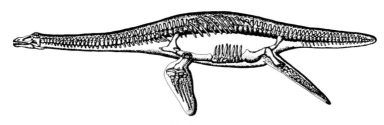

Plesiosaur

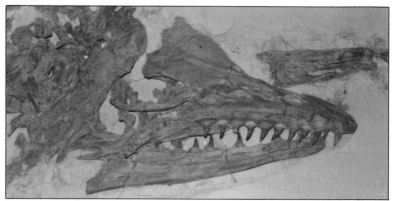

Mosasaur skull. Collection U of A.

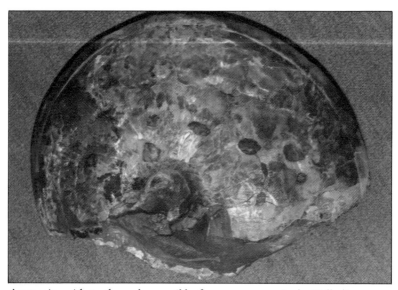

Ammonite with tooth marks, possibly from mosasaur attack. Collection U of A.

aquatic setting.

Mosasaurs' diet consisted mainly of fish and shellfish such as ammonites. Fossil shells of ammonites displaying tooth marks from mosasaurs are very common.

Mosasaurs survived until the end of the Cretaceous, 65 million years ago, but died out during the great extinctions that closed the Age of the Dinosaurs.

PART III

The Mystery Continues

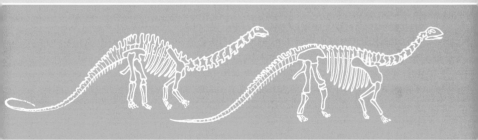

Dinosaur Eggs and Young

It is safe to say that the egg made the reptiles what they were. The eggs of reptiles were the first to develop hard or leathery outer shells unlike the eggs of amphibians, which had to be laid in water, or at least in moist terrestrial conditions. This significant development forever freed the reptiles, from an aquatic setting, thereby permitting them to exploit the land.

One theory suggests that during the Mississippian Period selective pressure among amphibians forced certain types away from wet regions and into drier areas. Here the jelly-like eggs of amphibians dried out, resulting in the destruction of the embryos. Over time, however, an egg with a hard shell evolved. This permitted the embryo to breathe, but prevented the loss of moisture from within. A special membrane called the *chorion* also evolved and allowed oxygen to pass through to the embryo and waste carbon dioxide to escape. The eggs contained a food source for the embryo and a liquid-filled sac (*amnion*) which acted as a shock absorber to protect the young reptile.

The eggs of reptiles, like those of birds, are fertilized within the mother's body before the hard mineral shell is deposited around them. Like modern bird eggs, the shells must be strong enough to prevent breaking but must be porous enough for gases to pass through, and more importantly, thin enough for the young embryo to break free during hatching.

Fossil eggs and young of *Protoceratops* were found during an expedition to Bayn-Dzak ("Flaming Cliffs") in the Gobi Desert during the late 1920s. These finds were astonishing. An entire group of dinosaurs was found in the vicinity of the nests: eggs, newly-hatched young, juveniles and adults. This find provided the first clue to the possible nesting habits of dinosaurs. Found in bowl-shaped depressions, the eggs were laid in a circle, and pointed outward like the spokes in a wheel. The eggs were laid several layers deep, suggesting they were covered with earth and left to hatch. The eggs were 20 centimetres long and sausage shaped. The shells were about 1 millimetre thick. (By way of comparison, the eggs of the modern ostrich are about 15 centimetres long.)

In 1978, the fossilized remains of 15 tiny dinosaurs were found

Dinosaur egg site at Devil's Coulee.

around a dome-shaped nest in Montana. The dinosaurs were surrounded by broken egg shells. These fossils have been indentified as belonging to a type of hadrosaur called *Maiasaura*. While an adult animal ranged from 7 to 9 metres long, these juveniles were only 1 metre long. The mound of earth which comprised the nest appeared to have been built for the protection of the tiny dinosaurs. These eggshell-strewn nests were bowl-shaped depressions about 2 metres wide and about .75 metres deep.

Subsequent excavations revealed more nests of *Maiasaura*. Six nests were found unoccupied, containing only fragments of eggshell, while one nest contained the fossil remains of even smaller hatchlings 50 centimetres long.

Because juveniles and hatchlings were found within the nesting colony site, this may indicate a degree of parental care. Tooth wear suggests that the juveniles fed on vegetation before they died. Perhaps this food was brought to the nest by adult *Maiasaura* or the youngsters fed in the vicinity of the nesting colony, perhaps under parental supervision. The presence of several adults likely deterred all but the largest predators from approaching the young.

Fossil evidence suggests that some disaster befell the adults responsible for this particular brood. Perhaps they were killed by predators, dooming the young to death by starvation — as

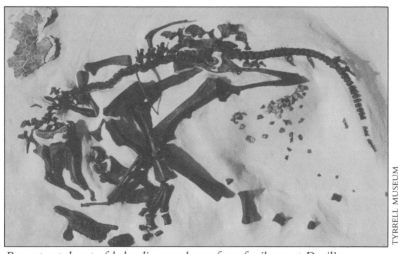

Reconstructed cast of baby dinosaur bones from fossil egg at Devil's Coulee. The length is about 30–35 cm. Tyrrell Museum.

their instinct to remain in the protective enclosure of the nest may have been very strong.

Protection of the young by adults is found in modern-day crocodiles, where the mother waits for the buried eggs to hatch before digging the young crocodiles out of nests and transferring them to safe areas called nursery pools. Here they are fed and protected from predators. Such behaviour may have been present among certain types of dinosaur.

In the summer of 1987, a spectacular find of hadrosaur eggs and hatchlings was made at Devil's Coulee, near Milk River, Alberta, by Kevin Aulenback of the Tyrrell Museum. The eggs, slightly larger than grapefruits, were found in a layer of grey sediment. The bones of the embryos can be seen within the fossil eggs as thin, white outlines about the thickness of a pencil. What is unusual about these finds is that the eggs are still in their original position in the nest. The position of the embryo bones seems to indicate that the eggs were ready to hatch but some factor, perhaps foul weather, prevented this. The nests are stacked in layers with about a half-meter of sandy clay between the nests, suggesting the area was used as a nesting site for several years in succession. The Milk River excavations are still very preliminary, being conducted by the Tyrrell Museum, Drumheller, Alberta, under the supervision of Dr. Phillip Currie.

Fragment of dinosaur eggs from Devil's Coulee.

Fragment of baby dinosaur jawbones (note tiny teeth).

More bone fragment from Devil's Coulee (note dollar coin for scale).

Warm-Blooded or Cold-Blooded?

One of the great debates of modern paleontology is whether the dinosaurs were warm-blooded or cold-blooded. For many years, the accepted theory was that, like modern-day reptiles, the dinosaurs were cold-blooded.

What is meant by "cold-blooded"? It means that the internal temperature of the animals (called ectotherms; ecto = external and therm = temperature) rises and falls as the surrounding temperature of the animal's environment changes. An ectotherm cannot regulate its body temperature internally, and as a result, cold-blooded animals are active mainly during the day, when the air temperature goes up. At night, when the temperature falls, the animal becomes very sluggish and remains inactive. This is the life pattern of modern reptiles.

Warm-blooded animals (called endotherms; endo = internal and therm = temperature) maintain a constant body temperature independent of the surrounding air temperature. This means a warm-blooded animal remains active at night when cold-blooded animals are inactive. The way an endotherm regulates its body temperatures is very complicated and has a great deal to do with the differences in the metabolic chemistry of warm-blooded animals as compared to the reptiles. Several adaptations of the warm-blooded animals, such as body hair in animals and feathers in birds, allow them to retain heat even when the surrounding environment is cold. Adaptations to reduce body temperature when the surrounding air temperature becomes higher than the ectotherm's internal temperature include sweating and panting.

Warm-blooded animals have several advantages over cold-blooded animals. Their ability to produce and regulate their internal temperature provides them with a far greater range of territory in which to live. Modern mammals live in all niches of the planet, from the high Arctic to the tropical rainforests, whereas modern reptiles are confined mainly to tropical and subtropical regions, where they compete quite successfully with endotherms.

Being warm-blooded has its price, however. To produce heat, the muscles of a warm-blooded creature must burn calories — a calorie is a measure of heat. One "calorie" when referring to food is actually one kilocalorie. One food calorie will raise the temperature of one litre of water one degree Centigrade. One

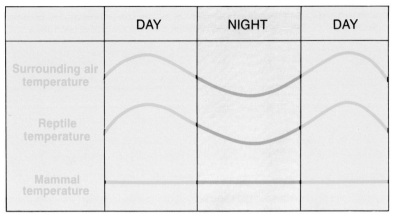

Chart shows the fluctuation of temperature between night and day. While the body temperature of the cold-blooded reptile is in steps with the temperature of the air, the internal temperature of the warm-blooded mammal remains constant.

hundred calories of food can provide enough energy to raise the temperature of one litre of water 100 degrees Centigrade. As you can see by this example, a small amount of food can produce a great amount of heat. In turn, this heat generates the internal temperature of warm-blooded animals.

Because the muscles of warm-blooded animals use calories to produce heat, the animal must eat ten times as much food as a cold-blooded animal of the same weight over the same period of time. A cold-blooded reptile would be much better suited to survive periods of scarce food than would a lion or tiger of the same weight.

Some paleontologists believe that a number of dinosaurs were warm-blooded like modern-day birds and mammals, and that the traditional view of the dinosaur as being a slow-moving creature living in swamps may be incorrect. Is there evidence to support the theory of warm-blooded dinosaurs? Were they quick and agile?

The answers may be found in the fossil bones of the dinosaurs themselves. Sections of dinosaur bone are ground very thin and examined under a microscope. Close examination reveals similarities between the structure of dinosaur bones and the bones of small modern mammals. Both the bones of dinosaurs and modern birds and mammals have large numbers of thin channels, called haversian canals, which carry blood through the bone. In

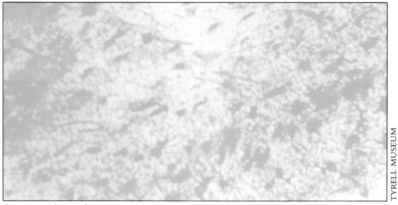

Dinosaur bone thin section.

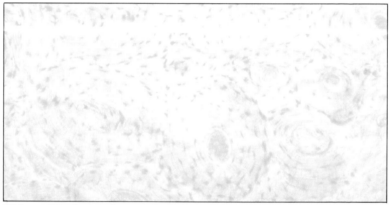

Thin section, human bone.

the bones of modern reptiles, turtles and snakes, fewer canals are found than in dinosaur and mammal bones.

While the bones of certain mammals and dinosaurs appear similar in structure, scientists cannot agree if these similarities prove that the dinosaurs were also warm-blooded. Other features of fossil bone and teeth are being studied for possible clues.

Some time may pass before paleontologists are able to agree on whether the fossil evidence supports a warm-blooded or cold-blooded theory, or both. Perhaps certain small theropod dinosaurs (which are thought to have been very active during life) were in fact warm-blooded, while larger sauropod dinosaurs like *Apatosaurus* were cold-blooded. We may never know for certain.

Death of the Dinosaurs

What caused the death of dinosaurs? Without a doubt, this is one of the most perplexing questions still facing paleontology today.

The dinosaurs were immensely successful, occupying all of the major terrestrial ecological niches, and dominating the earth for 160 million years. Yet, in the space of less than a million years, the entire dinosaur hierarchy was wiped out. While the theories attempting to explain extinction are many and varied, no one theory can stand up to all the contradictory evidence left by the collapse of the dinosaurs' reign.

Past theories ranged from hunters from outer space who killed off the great herds of dinosaurs to mammals eating the eggs of the dinosaurs. A discussion of the most popular theories will be of assistance.

Extraterrestrial Impact (Meteorite)

One of the most popular theories today is that of meteoric impact, proposed by Luis and Walter Alvarez of the University of California, Berkely. They stated that a meteorite of approximately 10 kilometres in diameter smashed into the earth. The collision and impact would have thrown great quantities of material into the atmosphere, blocking out the sun and triggering a global freeze not unlike a nuclear winter. The resulting drop in global temperature would have resulted in mass plant extinctions. Dinosaurs that relied on the plants for their survival would also die out. In turn, this collapse of the food chain would doom the predators which fed upon them.

Evidence to support this theory is found in a layer of clays composing the boundary between the Cretaceous and the Tertiary periods (KT boundary). Some paleontologists speculate that this is the line marking the extinction of the dinosaurs. In the clays, an element called Iridium has been found in concentrations which are much higher than those found in normal surface rocks. While iridium is very rare on the surface of the earth, it is 10,000 times more abundant in meteorites. Both the layers above and below this KT boundary layer are very low in iridium. Whatever happened at this boundary released great quantities of iridium into the sediments. This is not an isolated phenomenon, sampling confirms that the iridium anomaly seems to be worldwide and extends even into sea sediments formed at the same time.

Evidence found in ocean sediments seems to add weight to the extraterrestrial origin of the iridium. Organic molecules which are very rare on earth but quite common in some meteorites, called carbonaceous chondrites, were found mixed with the clay layer containing the iridium.

Recently, scientists have found small particles of shocked quartz (normally associated with meteorite impact sites) in a number of samples of boundary clays. Shocked quartz is the mineral silica which has been subjected to great pressure. When examined under a microscope, crossed lines running through the crystal structure become visible. The shocked quartz samples found in the boundary clay were subjected to pressures of over one million pounds per square inch. An estimate of the amount of energy a 10 kilometre meteorite hitting the earth would produce is on the order of 100 trillion tons of TNT.

While shocked quartz is sometimes found in violent volcanic eruptions, the markings in the grains are different from those found in impact sites having been exposed to less pressure than the boundary quartz. While there is no agreement on what these new findings mean, it seems likely that they may add support to the extraterrestrial impact which may have led to the extinction of the dinosaurs.

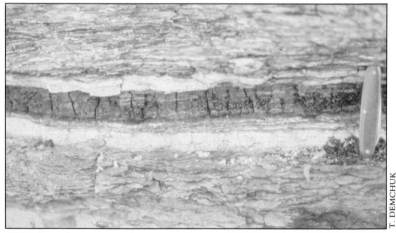

K.T. Boundary.

Volcanic Eruptions

Some scientists think that the extinction of the dinosaurs was a slow and gradual process triggered by an increase in global volcanic activity. This would have increased the levels of CO_2 in the atmosphere and caused a rise in global temperatures due to the greenhouse effect. Carbon dioxide slows the re-radiation of excess heat from the earth, thus causing a warming trend.

A rise in temperature affects the eggshells of birds; they become thinner and more likely to break. Perhaps the eggs of dinosaurs were also susceptible to this.

Dr. Heinrich Erben of Bonn University's Institute of Paleontology has been working in the French Pyrenees examining the eggshells and nests of dinosaurs in Late Cretaceous rocks just below the KT boundary. The results are astonishing! Eggshells from the lower, older layers were thicker-shelled (about 2.5 millimetres average for large eggs). But if the eggshells in the younger strata are examined, the shells grow progressively thinner. In the youngest rocks, near the KT boundary, the shells are only 1 millimetre thick. These shells would have been very fragile and easily broken.

In 1973, a nest of dinosaur eggs was found in Late Cretaceous rocks near Corbieres, France. The eggs were approximately 22.5 centimetres long by 17.5 centimetres around. Of the eight eggs in the nest, six were broken and two were still intact. When

Volcano, Phillipines.

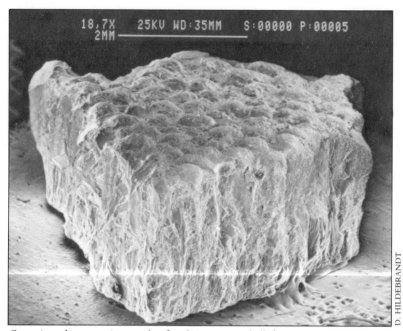

Scanning electron micrograph of a dinosaur eggshell fragment. Some fragments show a deficiency of calcium which may be linked to the death of the dinosaurs. Instruments like the S.E.M. (Scanning Electron Microscope) are important tools which allow scientists to determine elemental analysis of very small samples.

examined under a microscope, the shells were found to be very deficient in calcium. These shells were so poorly mineralized that the embryos could not have developed from them. Whether this was caused by stress or simply that the calcium in the fossil eggshells had been leached out during their burial is impossible to say. However, a simple rise in mortality of the eggs over several hundred generations would effectively wipe out large segments of a dinosaur population.

Supernova

One theory suggests that the explosion of a supernova near the earth exposed the earth to sufficient radiation to exterminate the dinosaurs and large parts of the biosphere. Dr. Dale Russell of the National Museum of Natural Sciences in Ottawa has suggested that perhaps it was not the radiation which killed the

Supernova Shelton 1987A. This exploding star is 163,000 light years away from earth. Some scientists theorize that such an explosion may have caused the extinction of the dinosaurs 65 million years ago.

dinosaurs directly, but the effects on the climate such a supernova would trigger.

Rise in Global Temperature

Paleontologists speculate that an increase in the global temperature may have caused the death of the dinosaurs. This increase may have been insufficient to be recorded in the geological record, but over time, would have been enough to stress the dinosaurs into extinction. Possibly a period of heightened solar activity caused the increase. A study of Upper Cretaceous oceanic plankton fossils reveals a continuous ocean temperature rise from the Jurassic to the Upper Cretaceous — the temperature then begins to decline. This increase in temperature may have caused the effects of thinning eggshells in Upper Cretaceous dinosaurs, thereby causing their extinction.

Changes in Vegetation

Perhaps a change in the type of plant life which evolved toward the end of the Cretaceous Period poisoned the dinosaurs. The angiosperms (plants with covered seeds; modern flowering plants such as oak, maple, poplar) evolved toward the end of the period. Perhaps the powerful alkaloids found in flowering plants poisoned the herbivorous dinosaurs while smaller mammals survived the poisonings due to changes in their biochemistry.

This theory does not enjoy wide support among scientists. There have been no ill effects noted on herbivorous lizards eating the same general type of vegetation. Therefore, it is highly suspect that angiosperms were responsible for the demise of the dinosaurs.

Magnetic Field Reversal

The magnetic field of the earth acts as a shield to filter out harmful cosmic rays. It is just possible that if the magnetic field were to reverse or become weak over long periods of time, the dinosaurs would have been exposed to increasing amounts of cosmic radiation which may have affected their reproductive success, thereby leading to their extinction.

Disease

Another theory suggests that the dinosaurs fell victim to a catastrophic plague which affected only the large reptiles. However, it is very rare that any one disease wipes out entire populations of one species, let alone broad families of animals. There is little evidence to support this theory.

Comet

Did a large comet collide with earth? One theory suggests as much. The impact of a large comet the size of Halley's could have set off a series of nuclear reactions, raising the global temperature. It should be noted that some comets contain significant amounts of cyanide compounds; perhaps this cyanide released into the atmosphere would have caused the Cretaceous extinctions.

Rise of the Mammals

The gradual rise of mammals in the Late Cretaceous, and their competing with dinosaurs, plus changes in climate and vegetation may have signalled the slow decline of the dinosaurs.

Fossil evidence found at Hell Creek and Bug Creek in Montana may provide a clue. An examination of the KT boundary at these sites, reveals the Iridium rich clays. The number of tiny mammal fossils found in ant piles seems to indicate an increase in the number of mammals as the KT boundary is reached. The last dinosaur fossils are found about 3 metres below this level, which seems to indicate that the dinosaurs were already extinct by the time the Iridium clays were deposited.

The answer to the extinction of the dinosaurs may be more complex and involve a number of factors. It is just possible that increased predation, by mammals or other species of dinosaurs, might also have decreased survival rates of eggs or juvenile dinosaurs. These factors, taken over one or two generations, would have had little effect on the populations, but even a small percentage rise in mortality, taken over several thousand years,

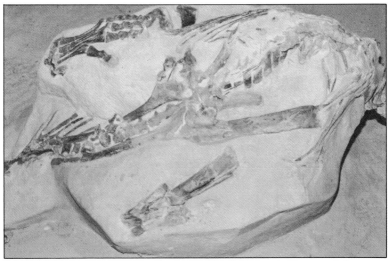

Montanoceratops. A relative of Triceratops, it was probably one of the last dinosaurs to become extinct in Alberta. It is thought that it lived in harsh enviornments where it had little competition from other dinosaurs. Fossil remains found in Crowsnest Pass, Alberta. Tyrrell Museum.

could have resulted in the death of whole populations of dinosaurs. The collapse of one part of the ecosystem would have resulted in greater pressure on surviving populations of dinosaurs, perhaps leading to a total collapse of the dinosaur ecosystem. This could have precipitated the extinction of the remaining dinosaur populations.

At best, the evidence that paleontologist have to answer these complex questions is fragmentary and contradictory. Few paleontologists agree on any one theory to explain the extinctions. Perhaps that is one of the reasons these giant dragons from our past hold such a special interest for us. After ruling the earth for 160 million years, their death and disappearance is still one of the great mysteries of science.

Plants of the Cretaceous

The plants of the Cretaceous were, like modern-day plants, the basis of the food chain. The plants growing during this period evolved rapidly and it is perhaps more than just a coincidence that the rise of the angiosperms or flowering plants coincides with the disappearance of the dinosaurs. While there is no evidence to support the theory, it has been suggested that the dominance of flowering plants in the Late Cretaceous may have contributed to the extinction of the dinosaurs.

The forests and lowlands of the Horseshoe Canyon Formation, typical of the area which is preserved in the rocks of the badlands of Drumheller, will serve as our example.

Until the appearance of the flowering plants in the Late Cretaceous, it is thought the primary food plants of the dinosaurs had been the ferns, conifers, cycads and horsetails. While it is difficult to state exactly what type of plants a particular dinosaur ate, science attempts to generalize the particular plant types based on the terrain or habitat that each type of dinosaur is thought to have occupied. Fossil evidence of particular plant types is very rare, but some clue may be found in fossil mummies and coprolites

Coprolite (life size).

(petrified dung). However, the difficulty with coprolites is that it is not possible to state what type of animal produced a particular coprolite.

Coprolites, when examined under a microscope, reveal a great deal of information about the diet and habitat of the particular animal from which it came. Certain micro-fossils may be found, fossil pollen can give clues to plant types, and if small bones are found, it may be possible to identify which types they came from. For example, the presence of fish scales would suggest that this particular animal ate fish. If small mammal bones are detected, it seems certain that the animal in question preyed upon small mammals.

If coprolites are examined for plant remains, the most likely candidates for identification are pollen grains. Should grains of pollen from conifers be found, it makes a strong case for the animal in question feeding upon conifers. With this information paleontologists conclude that, because this particular animal fed upon conifers, perhaps the type of terrain it lived in was similar to modern-day conifer forests.

While it is not possible to tell if a coprolite was produced by a dinosaur, these fossils do give scientists valuable clues to help them interpret the habitats in which the dinosaurs lived. And, as most mammals in the Late Cretaceous were very small, it seems a safe presumption that coprolites which are larger than these small mammals were probably produced by dinosaurs, or large reptiles. An examination of the fossilized stomach of a hadrosaur mummy revealed the presence of conifer needles. This evidence led some scientists to conclude that the hadrosaur had been feeding on conifer or metasequoia in the lowland forests before it died. This was in sharp contrast to the prevailing opinion that hadrosaurs were swamp dwellers. To illustrate how fossil evidence may be interpreted to give widely conflicting views, other scientists are of the opinion that the conifer needles may have simply washed into the rotting carcass of the dead dinosaur, and bear no relationship to its diet. There is no way to tell if either of these views is correct.

The difficulty in piecing together ancient life from fossils is that the information is so incomplete, much like trying to reconstruct a play by Shakespeare from small fragments. Imagine digging up a book of his complete works from a garbage dump. Even if you recover hundreds of fragments, each so tiny that it contains only a single word, it will not be possible to reconstruct the entire play. Certain parts will be jumbled, while whole scenes may be missing or were destroyed. This will give you some idea of how difficult it is to reconstruct the life of the dinosaurs. But the problem is even more complex than the book of plays. Imagine that, instead of one book by Shakespeare, you had buried a whole library in the dump. You had no idea about what books were buried, or even how many. Now, if you recover tiny scraps of paper, do they come from a book of plays, a telephone book or a line from the classified ads? Perhaps this example illustrates the difficulties faced by paleontologists in reconstructing the world of the dinosaurs.

The following plants are typical of those plants dinosaurs would have eaten during the Late Cretaceous. Most are found as fossils in the Horseshoe Canyon Formation around Drumheller.

Cycads

Resembling palm trees, cycads had short, stout trunks. Cycads were Gymnosperms and fertilized their seeds with pollen. Like conifers, cycads produced woody cones. Very common during the Mesozoic, cycads are thought to have evolved in the Permian about 250 million years ago. Very likely they were used as food by dinosaurs right through the Triassic to the Cretaceous. They throve in warm, moist areas such as jungles or marshlands. Their fossils are found in the rocks of the Upper Cretaceous.

Ginkgo

Only one species of ginkgo still survives today. Native to China, ginkgo trees were removed from the wild centuries ago and tended by monks. Like the conifers and cycads, the ginkgos produce pollen and display characteristics common to both conifers and cycads. Ginkgos are thought to have evolved in the late Triassic or early Jurassic about 200 million years ago. The habitat of the ginkgos is cooler and dryer than that of the cycads. They may have been good plants for those dinosaurs inhabiting the upland forests. The fossil leaves of ginkgos are very common in the rocks of the Upper Cretaceous.

Cycad

Ginkgo

Araucaria, a living fossil.   *Living Metasequoia.*

Fossil impression of Metasequoia, Cretaceous. Alberta.

Araucaria

These plants are thought to be survivors of plants which evolved about 225 million years ago. Producing cones like conifers, these cones are commonly found as fossils in Mesozoic rocks. Like modern pines, the probable habitat of Araucaria was upland or lowland forest, drier than the habitats of ferns and horsetails, which favoured the swamps. While there is no fossil evidence to indicate which type of dinosaurs, if any, fed upon Araucaria, they were very common and grew in great profusion. Therefore, it is probable that some dinosaurs incorporated Araucaria into their diet — perhaps sauropod dinosaurs browsed in the Araucaria high branches. Fossils of Araucaria are commonly found in Upper Cretaceous rocks.

Metasequoia

The fossil impressions of Metasequoia are found in the rocks of the Horseshoe Canyon Formation. Related to the Sequoia, this ancient cone-bearing gymnosperm was found living in China in 1948. Metasequoia is unique in that its fossilized remains have been found in the fossilized stomach of a hadrosaur. The living plant today seems unchanged from its ancestors which flourished in the late Cretaceous 65 million years ago.

These are only a few of the plants which flourished during the Age of the Dinosaurs. Today's forests are filled with living reminders of that vanished age. Oak, poplar and chestnut trees were surely food for some reptilian palate. The more we search the fossil record, the more evidence we find for the diversity of plant life which sustained these giants.

How Old Is It?

One of the most common questions asked by people interested in dinosaurs is, "How old is it?" This question is always followed closely by, "How do you know?" There are a number of ways to answer these questions. One method is called radiometric dating and the other is relative dating.

Radiometric Dating

One method scientists use to determine the age of a rock involves the decay of certain unstable radioactive elements. This method can be used to date certain volcanic rocks. Sometimes in a *rock sequence* (series of layers of rocks) we find layers of ancient volcanic ash. This material is very common in the badlands of Alberta, and can be used to give us a date.

Let us say that a dinosaur fossil we are trying to date was deposited in a layer of volcanic ash. Perhaps the dinosaur was killed and buried by a cloud of ash when a volcano erupted. This is not as strange as it may sound: several years ago, when Mount St. Helen's erupted, the ash cloud was carried over several states and parts of British Columbia and Alberta. This ash settled into lakes and streams, killing many fish and animals, and their remains became part of the ash sediments which were deposited the year of the eruption.

During the Age of the Dinosaurs, many volcanoes were erupting and ash clouds were carried across the land. Some of these ash beds contain dinosaur bones. The layer of ash allows us to date the dinosaur fossil. Because the volcanic ash contains radioactive atoms of Potassium 40, scientists use a method known as Potassium-Argon dating to determine the age of the ash.

The radioactive element Potassium 40 is found in certain minerals in the volcanic ash. Over time, this Potassium 40 breaks down into Argon 40. Argon 40 is a very rare element, and any that we find in a rock is certainly the result of the radioactive decay of Potassium 40. In 1.3 billion years, one-half of all the Potassium 40 in the rock will have become Argon 40.

By melting the rock sample, it is possible to drive off the Argon gas and to determine the age of the rock by the amount of Argon which is produced.

In the case of our dinosaur bone, if the age of the volcanic ash which surrounds it is 70 million years, we can state that the age of the dinosaur bone is about 70 million years.

Usually several methods of radiometric dating are used, along with other means of dating the rock strata, to arrive at a true age of the sample. Of course, this is a very simple explanation of just one method of radiometric age determination, but it shows you how scientists can date a layer of rock.

Relative Dating

Paleontologists compare fossils from one area with those from another, and by using the fossils themselves, it is sometimes possible to estimate the age of a rock. For example, let us say you have a rock which you are trying to date. You carefully examine the rock and find that it is limestone and was deposited in the ocean. There are two types of fossil shells in the rock, which were living together in the mud on that ocean floor. We will call these Fossil A and Fossil B.

By consulting information gathered by other paleontologists from around the world, you learn that shells of the same type as Fossil A have been found in ash beds around the world. When these ash beds were dated using the Potassium-Argon method, none of the shells were found to be younger than 400 million years old.

Shells of the same type as Fossil B have also been found in ash beds, dated by the Potassium-Argon method and never found to be older than 425 million years.

Because our unknown rock contains Fossil A, we know that it must be older than 400 million years old, because no rock containing Fossil A is younger than 400 million years. This tells us that our rock is older than 400 million years, but how much older?

Because Fossil A and Fossil B were deposited in the same rock, they must have been living together at the same time. We know from our Potassium-Argon dates that rocks containing Fossil B have never been older than 425 million years.

This means that our rock is not older than 425 million years, because of Fossil B, and not younger than 400 million years, because of Fossil A. Therefore, our unknown rock is between 400 and 425 million years old. This explanation is, of course, very much simplified, but I hope it gives you some insight into the methods paleontologists use to date fossils.

Dinosaur Provincial Park

Dinosaur Provincial Park was established in 1955 to protect western Canada's unique dinosaur fossils. It has been declared a UNESCO World Heritage Site. The park is situated northeast of Brooks, Alberta and encompasses 90 square kilometres of badlands and prairie along the Red Deer River. The fossil remains of Cretaceous dinosaurs, along with primitive mammals, birds, reptiles, amphibians and marine organisms can be found within the park.

This unique fossil assemblage has come about as the result of a coastal flood plain which formed along the eastern flanks of the Rocky Mountains about 80 million years ago. The rich marsh environment, which resembled the modern coastal regions of the southern United States, such as Louisiana, supported a wide variety of now-extinct plants and animals. The sediments laid down in this environment seemed to favour the preservation of fossil materials, many of which may be seen today eroding out of the badlands of the park.

The term "badlands" was first used because such lands were useless for cultivation, due to the steep hoodoos and coulees characteristic of the region. This topography is the result of erosional processes which have taken place since the last Ice Age, about 10,000 years ago. A large number of the sediments which make up the badlands contain bentonite, a volcanic ash which expands when exposed to water. It is the presence of these expanding clays which allows the formation of the characteristic steep-sided hoodoos and coulees common throughout Dinosaur Provincial Park.

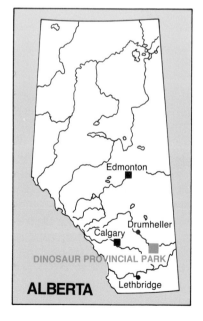

View of Dinosaur Provincial Park.

Certain areas of the park are restricted to public access to allow scientific work to be conducted. During the summer months, hiking and bus tours are conducted. Among the other features of interest within the park are the dinosaur field digs conducted most summers by the Tyrrell Museum of Paleontology, which is located in Drumheller, Alberta. If you would like more information concerning Dinosaur Provincial Park, contact the Alberta Department of Tourism.

Pioneer Dinosaur Hunters in Canada

The history of dinosaur collecting in Canada begins with the discovery of hadrosaurian bones by George M. Dawson, during his work with the International Boundary Commission (1873-75), while surveying the 49th Parallel. The remains were submitted for identification to E. D. Cope, the Philadelphia naturalist and famous U.S. dinosaur hunter. The following year, dinosaur remains were found in Southern Alberta in the Belly River beds.

Later expeditions in 1884 brought Joseph Burr Tyrrell, an assistant of Dawson from the Geological Survey of Canada, to southern Alberta. Here, in the valley of the Red Deer River the first large, well-preserved skull of a carnivorous dinosaur, was found. This fossil skull was sent to Calgary by wagon and finally to Ottawa. It was named *Albertosaurus sarcophagus*.

Discoveries near Steveville, a small settlement on the Red Deer River, brought other collectors. In 1897, Lawrence M. Lambe, a geologist from the Geological Survey came to Alberta. (*Lambeosaurus* was named for him.) L.M. Lambe worked closely with the American Museum of Natural History in New York. Lambe's monographs on dinosaurs of southern Alberta alerted the world to the proliferation of rich fossil beds in the Edmonton and Oldman Formations. The news brought collectors from around the world, most notably the United States, and ushered in the golden age of dinosaur collecting in Canada along the Red Deer River.

The Sternbergs (C.H. Sternberg and his son, C.M. Sternberg) played a very large part in the history of dinosaur collecting in Canada. C. H. Sternberg had worked with the famous dinosaur collector, Cope, in Kansas. Sternberg and Cope also collected dinosaurs in Montana. C. H. Sternberg and his sons, Charles and Levi, became contract fossil collectors, working for most major museums. Charles and Levi later came north, leaving the United States to work for the Geological Survey of Canada.

Barnum Brown, born in 1873, worked for the American Museum of Natural History in New York. He was one of their greatest collectors, having journeyed around the world to bring some of the great fossil dinosaurs back to the AMNH. It is likely that no one individual collected more dinosaurs in his lifetime

George M. Dawson. Collection U of A.

Joseph B. Tyrrell.

than Barnum Brown. In 1902, he helped collect the finest specimen ever found of *Tyrannosaurus rex* in the Hell Creek Formation of eastern Montana. In 1910, Barnum Brown came to Alberta, alerted to the rich dinosaur beds by the monographs of Lambe and by a rancher, John Wagner, who visited the museum in New York and remarked on the similarity between fossils on display and fossil bones on his land along the Red Deer River.

The spectacular finds that Barnum Brown made along the Red Deer River and the large number of fossils being removed to museums in the United States prompted the Geological Survey of Canada to fund its own expeditions to the badlands of Alberta. The Sternbergs, hired to collect dinosaur fossils for the Survey, collected in Canada for several seasons. The results of the finds included hadrosaur fossils of *Corythosaurus*, horned dinosaurs, and *Chasmosaurus* complete with skin impressions.

Charles M. Sternberg worked for the Geological Survey of Canada most of his life, while his brother Levi joined the staff of the Royal Ontario Museum in 1919. In 1921, Charles Sternberg was hired by the University of Alberta to collect a fossil *Corythosaurus* from the Red Deer River Valley. The specimen is now housed in the paleontology museum in the Earth Sciences Building on the campus of the University of Alberta in Edmonton, Alberta. The pictures of this expedition are supplied by the University of Alberta archives and show the primitive conditions under which some of these fossils were collected.

Party members working on in situ fossils, 1921 Sternberg Expedition.

Moving fossils from canyon with team of horses. 1921 Sternberg Expedition.

Party members wrapping fossil in plaster bandages for shipment. The gentleman on the right is probably G.M. Sternberg. 1921 Sternberg Expedition.

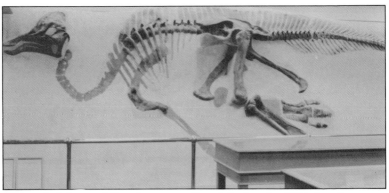

Mounted skeleton of Corythosaurus collected during the 1921 Sternberg expedition. Collection U of A.

Dr. John A. Allan of the University of Alberta was instrumental in obtaining the Sternberg dinosaur collection for the University of Alberta in 1934 - 1935. As head of the Department of Geology for 29 years, Dr. Allan was largely responsible for the creation of the University of Alberta's extensive collection of fossils, which are now used for study by paleontologists from around the world. First appointed to the Department of Geology in 1912, Dr. Allan spent much time in the field, and was personally responsible for the collection of many dinosaur fossils, particularly a skull of Albertosaurus which is probably one of the finest known to science. He remained quite active in the field of geology until his death in 1955. His dedication over a lifetime contributed greatly to the understanding of the dinosaurs of Alberta.

Dr. Allan

Field Collecting

The process of collecting a dinosaur fossil in the field is more complicated than it first appears. When the bones were first buried, the minerals in solution in the groundwater, mainly SiO_2 (silica), permeated the soft bone tissue and hardened them. Over millions of years, the pressure of the overlying sediments and the later unloading by erosion resulted in the brittle mineralized bones being fractured into multiple fragments. For example, a single bone might be reduced to a mass of several hundred fragments.

The field collection of a dinosaur allows the paleontologist to remove as much material from the field environment to the lab, where the painstaking process of re-assembly can take place. The first step in collecting the dinosaur is a careful mapping of the site. All information such as location, type of formation, etc., are noted to allow a precise dating of the skeleton. The strata around the skeleton are searched and sifted for bone fragments which might have eroded from the main find. The amount of information

Block turned over to allow plastering of underside. Almost ready to transport.

obtained from this process is tremendous. Small bones such as teeth, or tiny fragments of skull bones or tendons can be recovered in this way. Other information can be obtained from fossil seeds or plant fragments which may be found in close proximity to the skeleton. This can provide clues to the environment in which the dinosaur lived and died. All this information is carefully mapped and recorded so that when the site is mapped, the location of each and every find can be plotted. Each stage in this collection process is carefully photographed to provide a complete record, which allows the paleontologists to reconstruct the fossil dinosaur.

After the rock or earth is removed from around the dinosaur skeleton, it is sifted and searched for fossils. A thin layer (5 to 10 centimetres) of rock or earth is left around the fossil skeleton to protect it during transport. Sometimes it is necessary to use shovels, bulldozers or dynamite to remove this overburden. In other cases, fine brushes and dental picks clean and prepare the fossil dinosaur for transport.

When the major portion of the overburden is removed, the skeleton is coated with layers of tissue paper. Then bandages soaked with plaster of Paris are wrapped around the fossil dinosaur

Moving the blocks with bulldozer. Compare this with the team of horses in 1921!

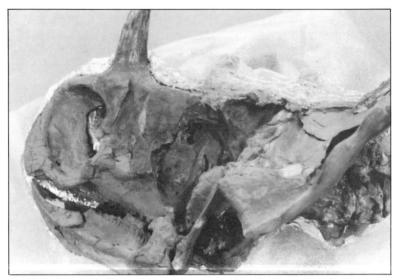

Partially prepared skull of Centrosaurus. Notice the plaster bandages which are used to protect the fossil during transport. Collections U of A.

to protect it during shipping. When the plaster dries rock-hard, the ground beneath the fossil dinosaur is dug away, leaving a 5 to 10 centimetre protective layer under the dinosaur. The fossil skeleton is turned over and the underside coated with paper and plaster bandages. This provides a very hard plaster cocoon which allows the fossil to be removed from the field. In some cases, the fossil skeleton might be small and can be removed in one block. With larger dinosaurs, the skeleton must be cut up into convenient-sized blocks for transport. Each of the blocks is given a number which is painted on the block, then the blocks of the fossil skeleton are transported to the lab either by helicopter or truck.

Once the skeleton reaches the lab, the process of reconstruction continues. The plaster jacket is first cut away from the skeleton and laboratory technicians begin the work of preparing the skeleton for study or display. Cleaning the bones can be very time-consuming; months or even years can be spent preparing a single skeleton. Fine dental tools, drills and chisels are used to clean the rock from around the bones. This work must be done very carefully, much can be learned about the dinosaur while it is being prepared. There are rare cases where the soft mud and sediments which once surrounded the body when buried preserved an

Partially excavated hadrosaur jaw. Note the interlocking batteries of teeth.

impression of its skin. This would be a very rare fossil and must be saved and displayed when the skeleton is prepared for study. Skin impressions are very rare, but sometimes found in the fossils of Hadrosaurs found in Dinosaur Provincial Park and in Wyoming. Careful preservation of these details provides scientists with even more information on the details of dinosaur anatomy.

Once the fossil skeleton is cleaned and prepared, the bone fragments must be glued together to form complete bones. This is a job for a very skilled person with much patience and a knowledge of anatomy. The gaps between the bone fragments are filled with plaster which is then painted to resemble the bone. Steel pins and supports may be used to provide strength for the fragile and brittle fossils if they are to be mounted in a free-standing skeleton. When this process is complete, the bones are sometimes treated with preservatives which prevent the bones from breaking or deteriorating further.

Following preparation, most skeletons are used for study or display. Sometimes a fibreglass cast is made of the fossil bones. This material is better than rock, because it is strong and light. A mold of each bone is made and used to create a fiberglass replica of the original. These copies are perfect in every way, preserving

Kevin Aulenback, who found the nests at Devil's Coulee.

Joshua Enookolook with hadrosaur toe bone from Bylot Island.

each tiny crack or detail. The casts are then painted to resemble the original bones. While many skeletons displayed in museums are composed of complete and original fossilized bone, in some cases, skeletons may be completed with bone casts. Sometimes the skeletons are actually casts made entirely from fiberglass.

By making copies of the bones, each of the museums contributing part of a skeleton will have a complete skeletal cast for their collection. The original bones are stored for study and kept safe from breakage. This ensures that if a cast skeleton is damaged or destroyed, another copy can be made from the original molds.

Some dinosaurs were very rare. In the entire world, there have been fewer than 10 skeletons of the *Tyrannosaurus rex* found. Yet there are hundreds of casts in museums around the world made from these bones.

The time between field collecting and museum display can be years. Each reconstruction represents years of work by dedicated individuals who have brought the great dinosaurs back to life from the dust of the earth.

Recent Dinosaur Find

In the summer of 1987, Dr. Elliott Burden and a research team from Memorial University of Newfoundland's Centre for Earth Resources Research found the first dinosaur bones ever discovered in the Canadian High Arctic. The bones, identified as fossil remains of a hadrosaur (duck-billed dinosaur), were found on Bylot Island, which lies off the northern coast of Baffin Island.

Besides Dr. Burden, the party consisted of Kerry Sparkes and James Waterfield, both Master of Science students at Memorial. The initial discovery of the bones was made by a fourth member of the party, field assistant Joshua Enookolook.

The purpose of the expedition was to collect samples of ancient pollen and spores. Along with the dinosaur fossils were found the remains of a mosasaur (marine reptile) and a great deal of other assorted fossil material.

The importance of this historic find cannot be overstated. Dinosaur fossils have never been found before in the High Arctic. While scientists have long speculated about dinosaur populations living in the region during the Cretaceous period, this historic find represents the first piece of physical evidence to support the theories, and to confirm that 65 to 75 million years ago, the Canadian High Arctic boasted a sub-tropical climate.

The Role of the Dedicated Amateur

Many people who are interested in dinosaurs often try their hand at field collecting. The idea of finding a rare skeleton of a new, previously-unknown dinosaur is a thrilling one. Both scientists and dedicated amateurs alike can contribute something to the study of dinosaurs.

Understandably, before a dinosaur can be excavated it must first be found. When a museum or university funds a dig, it spends thousands of dollars to put people into the field to look for dinosaurs. These people are highly trained and know what to look for and where to look for dinosaur fossils, but each person has only two eyes and cannot be everywhere. This is where the public can be of enormous help to science. For every researcher in the field, there are hundreds of rock collectors, fossil hunters, farmers, hunters, fishermen and outdoor enthusiasts. These people can function as the eyes of science. Many of the great dinosaur finds have been made by people who are not scientists. Rare dinosaur nests were found at Devil's Coulee in Southern Alberta, due to the efforts of Wendy Sloboda, a grade twelve student and amateur paleontologist, who was working as a volunteer for the Tyrrell Museum at the time. The most important thing to remember is that if you find an important fossil, **do not touch it or attempt to remove it from the ground. Call a museum or government agency, or the geology department at your local university.** This is very important. In this way, every citizen can contribute toward the study of dinosaurs. Perhaps your small find might be the single find which provides the final clue to the solution of one of the mysteries of the dinosaurs.

The study of fossils is an engaging hobby which will teach you about the world of today and yesterday and can take you into the outdoors for fresh air and excitement. The equipment you need for this hobby is simple and inexpensive:
- a 10 X magnifying glass (approximate cost $5.00)
- good, stout boots
- notebook, coloured pencils
- paper for wrapping samples
- camera to photograph fossils
- cards for cataloging your collection
- book for identification of fossils and rocks
- knapsack

Please remember that rock quarries and riverbanks where rocks and fossils are found are extremely dangerous. Never attempt to collect there unless you are under the supervision of a responsible adult.

There are a number of clubs and organizations which are interested in the study of fossils or rocks and minerals. Local boys' and girls' clubs, Scouts and Girl Guides might be able to provide information regarding activities in your local area. Rock and lapidary clubs or shops should also be able to provide information.

If you require information about regulations or information about fossil or rock hunting in your area of Alberta, I suggest you contact either the Tyrrell Museum of Paleontology in Drumheller, Alberta, or the Provincial Museum and Archives in Edmonton, Alberta. Additional helpful information might be obtained by contacting the Department of Geology (fossils and rocks) or Department of Zoology (fossils) at the University of Alberta or University of Calgary. In other provinces of Canada or areas of the United States, it is recommended that a collector get in touch with Provincial or State agencies to determine what local regulations apply to this hobby.

There are certain rules which must be followed when dealing with dinosaur finds:

In Alberta, it is illegal to dig for fossils without a permit from the government. There is a maximum fine of $50,000.00 for this offence. This applies if the fossils are found on private land, or more importantly, in a park. Any fossil found in a park must not be touched, moved or disturbed

in any way. It is permitted to photograph fossils in a park, providing they are not disturbed or moved in any way.

It is very important that you do not disturb any dinosaur fossils you find. Very important information may be destroyed if you attempt to excavate a fossil skeleton. This is a job which must be left to trained professionals.

If you are interested in digging for dinosaurs, there is a program which is offered most summers by the Tyrrell Museum of Paleontology in Drumheller. The program runs for several weeks and allows volunteers (all of whom must be over the age of 18) to excavate dinosaur fossils under the supervision of trained paleontologists. I suggest you write the Museum for information regarding this summer program.

If you are in an area which is not close to suitable collecting areas, or perhaps you are shut in and cannot get out, you may purchase fossils from dealers who supply catalogues by mail. This is an inexpensive way of building a collection of fossils which cannot be found in your area. You can obtain fragments of dinosaur bone, gastroliths, or coprolites which would normally be impossible for you to obtain any other way. I suggest you check your local telephone directory for listings of lapidary shops or mineral and fossil dealers in your area. For listings of foreign dealers, you may wish to consult foreign directories in your public library.

Pronunciation of Dinosaur Names

1. Brachiosaurus .... brack-ee-oh-SAW-rus
2. Diplodocus .... dip-LOD-oh-kus
3. Apatosaurus .... a-PAT-oh-SAW-rus
4. Camarosaurus .... kam-AR-a-SAW-rus
5. Protoceratops .... pro-toe-SER-a-tops
6. Iguanodon .... i-GWA-no-DON
7. Edmontosaurus .... ed-MONT-oh-SAW-rus
8. Maiasaura .... MY-a-SAW-ra
9. Stegosaurus .... STEG-oh-SAW-rus
10. Stegoceras .... steg-O-ser-as
11. Triceratops .... try-SER-a-tops
12. Centrosaurus .... SEN-tro-SAW-rus
13. Chasmosaurus .... KAZ-mo-SAW-rus
14. Ankylosaurus .... an-KY-lo-SAW-rus
15. Allosaurus .... AL-oh-SAW-rus
16. Tyrannosaurus .... tie-RAN-oh-SAW-rus
17. Albertosaurus .... al-BERT-oh-SAW-rus
18. Deinonychus .... DIE-no-NIKE-us
19. Struthiomimus .... STROOTH-ee-oh-MIME-us
20. Archaeopteryx .... ARC-ee-OP-TER-iks
21. Pterosaur (Pteranodon) .... TERR-an-O-don
22. Ichthyosaur .... ICK-thee-o-sore
23. Plesiosaur .... PLEASE-ee-o-sore
24. Mosasaur .... MOZZ-a-sore
25. Supersaurus .... super-SAW-rus
26. Ultrasaurus .... ULL-tra-SAW-rus
27. Corythosaurus .... ko-RITH-oh-SAW-rus
28. Parasaurolophus .... par-a-SAWR-oh-LOAF-us
29. Lambeosaurus .... LAM-bee-oh-SAW-rus
30. Kritosaurus .... KRIT-oh-SAW-rus
31. Anatosaurus .... an-AT-oh-SAW-rus

Glossary

A

ABSOLUTE DATING: A means of determining the age of a rock by using radioactive isotopes. See also *Radiometric Dating*.

AMMONITES: Extinct coiled shellfish related to *Nautilus*, squid and octopus.

ANAPSID: A group of reptiles related to living turtles and tortoises.

ANGIOSPERMS: Flowering plants.

ARTHROPODS: Animals with jointed legs, such as insects, spiders, trilobites (extinct), crabs and shrimp.

B

BACTERIA: Microscopically small organisms (plants and animals).

BIPED: An animal which habitually walks on two legs.

C

CALCITE: Mineral found in the Earth's crust ($CaCO_3$). Examples are limestone and chalk.

CAMBRIAN PERIOD: The earliest Paleozoic time period, 570 to 500 million years ago. Earliest rocks in which fossils are common.

COMPARATIVE DATING: A method used to estimate the age of rocks by the fossils they contain. Fossils from rocks of an unknown age are compared to fossils of a known age. This method allows scientists to determine the approximate age of a sample.

CONIFERS: Cone bearing trees, such as pine, spruce and fir.

COPROLITES: Fossilized dung or droppings.

CRETACEOUS PERIOD: Time period which extended from 140 million to 65 million years ago. The end of this period is marked by the extinction of the dinosaurs.

CYCADS: Plants which resemble squat palms, which were quite common during the time of the dinosaur.

D

DENTITION: Teeth.

DEPOSIT: Accumulation of rock or material (sediment).

DIAPSID: A reptile group related to crocodiles, lizards, snakes and birds. The dinosaurs were part of this group.

DINOSAUR: Land dwelling reptiles which walked erect, and flourished between 225 and 65 million years ago. The classification of dinosaurs is based on two orders: Saurischia and Ornithischia.

E

ECHINODERMS: Starfish and their relatives.

EMBRYO: The earliest stages of an animal's development.

ENDOTHERMIC: "Warm-blooded" animals able to regulate the temperature of their bodies by means of chemical reactions.

EVOLUTION: A gradual change in the characteristics of plants or animals over time, brought about by the process of natural selection.

EXTINCTION: The death of a species.

F

FLORA: Plants.

FOLIAGE: Leaves, branches and twigs of plants.

FOSSIL: The preserved remains of something which once lived.

FOSSILIZATION: The process by which fossils are produced.

G

GASTROLITHS: Stomach stones, thought to have been used as ballast or for grinding and pulverizing food.

GEOLOGICAL TIMESCALE: A timescale of Earth's history based on absolute and comparative dates of rocks and fossils.

GEOLOGIST: A person who studies rocks.

GEOLOGY: The study of rocks.

GINKO: A species of tree (Maidenhair) which is the sole survivor of a group of gymnosperms which flourished during the time of the dinosaurs.

I

ISOTOPE: Atoms of an element which differ in atomic weight. Some isotopes are more unstable than others and tend to break down. This is known as decay. Isotopes are used in radiometric dating.

J

JURASSIC PERIOD: A period of Earth's history which lasted from 200 to 135 million years ago.

L

LIGAMENTS: Strong sheets or threads of protein which support the joints between bones.

M

MINERALIZATION: The process whereby bone or wood is replaced by minerals dissolved in groundwater to produce a copy of the original. Some common minerals involved in this process are $CaCO_3$ (Calcium Carbonate, Limestone) and SiO_2 (Quartz).

P

PALEONTOLOGIST: A person who studies fossils.

PALEONTOLOGY: The study of fossils.

PETRIFICATION: "Turning to stone"; the replacement of organic tissue of a fossil with minerals.

PRECAMBRIAN: Period of time from Earth's formation (4.6 billion years ago) to 600 million years ago. Fossils are not common in Precambrian rocks.

PREDATORY: Preying on other animals by hunting and killing them.

Q

QUADRUPED: An animal which habitually walks on four feet.

R

RADIOACTIVE DECAY: The disintegration of atoms of unstable elements, producing other elements, and energy.

RADIOMETRIC: The measurement of radioactive decay.

RESONATOR: A device for increasing the intensity (loudness) of sound.

S

SEDIMENTS: Grains of sand, clay, mud, silt which are deposited, forming sedimentary rocks.

SEDIMENTARY ROCKS: Rocks formed by the deposition of sediments.

SHALES: Rocks formed from clay sediments, which have a tendency to split into flat plates like leaves in a book. Excellent for the preservation of fossils.

SILICA: (Quartz) Very common mineral (SiO_2). Grains of silica are the major component of sand.

SPECIES: A group of related animals that look alike and can interbreed.

STROMATOLITES: Banded rocks composed of calcium carbonate or limestone, which were produced by mounds of blue-green algae. Common in Precambrian times.

SUPERNOVA: The last stages in the life of a star, during which it explodes, producing great amounts of energy as it destroys itself.

T

TALON: Sharp claws.

TERRESTRIAL: On the Earth's surface; can also mean "land-dwelling".

THEROPODS: A range of Saurischian dinosaurs, mostly carnivores (meat-eaters). Small theropods are referred to as coelurosaurs (eg., Oviraptor), while large theropods, such as Albertosaurus, are referred to as carnosaurs.

TRIASSIC PERIOD: A period of time in Earth's history about 225 - 200 million years ago. Dinosaurs first appeared toward the end of the Triassic.

VERTEBRA: An individual bone of the back. All these bones taken together make up the backbone, or spine.

VERTEBRATES: Animals with backbones, or spines.

Museums

Dinosaur Provincial Park
Patricia
Alberta

**National Museum of
Natural Sciences**
Ottawa
Ontario K1A 0M8

**Provincial Museum
of Alberta**
12845 102 Avenue
Edmonton
Alberta T5N 0M6

Redpath Museum
McGill University
859 Sherbrook Street West
Quebec H3A 2K6

Royal Ontario Museum
Toronto
Ontario, M5S 2C6

**Tyrrell Museum of
Palaeontology**
PO Box 7500
Drumheller
Alberta

**American Museum of
Natural History**
Central Park West/79th St
New York, New York 10024

**Denver Museum of
Natural History**
City Park
Denver
Colorado 80205

Dinosaur National Monument
PO Box 128
Jensen
Utah 84035

**Earth Sciences Museum
Brigham Young University**
Provo,
Utah 84602

**University of Wyoming
Geological Museum**
Box 3254
Laramie
Wyoming 82071

**Utah Museum
of Natural History**
University of Utah
Salt Lake City
Utah 84112

The Author

Ron Stewart is a native of Edmonton, and works for the Department of Geology at the University of Alberta. One of his duties with the Department is conducting public lectures and tours of the dinosaur exhibits in the Department of Geology's Paleontology Museum. It was as a result of these talks that *Dinosaurs of the West* was written. Ron Stewart has been writing for eight years and has produced material for CBC and a number of newspaper articles discussing local history. He currently resides in Edmonton with his wife Michelle, and Muggins The Cat, who are both providing moral support as he completes forthcoming writing projects.

MICHELLE STEWART